Collins

Greater Depth Maths

Build curiosity and confidence

TEACHER GUIDE UPPER KEY STAGE 2

Herts for Learning

William Collins' dream of knowledge for all began with the publication of his first book in 1819. A self-educated mill worker, he not only enriched millions of lives, but also founded a flourishing publishing house. Today, staying true to this spirit, Collins books are packed with inspiration, innovation and practical expertise. They place you at the centre of a world of possibility and give you exactly what you need to explore it.

Collins. Freedom to teach.

Published by Collins
An imprint of HarperCollins*Publishers*
The News Building
1 London Bridge Street
London
SE1 9GF

HarperCollins*Publishers*
1st Floor, Watermarque Building
Ringsend Road
Dublin 4
Ireland

Browse the complete Collins catalogue at

www.collins.co.uk

10 9 8 7 6 5 4 3 2 1

ISBN 978-0-00-845493-7

British Library Cataloguing-in-Publication Data
A catalogue record for this publication is available from the British Library.

Authors: Charlie Harber, Rachael Brown, Nicola Adams, Laura Dell
Publisher: Lizzie Catford
Product manager: Letitia Luff
Editorial management: Oriel Square
Cover designer: Amparo Barrera
Cover photograph: Sasirin Pamai/Shutterstock
Typesetter: Jouve India Pvt. Ltd.
Internal illustrations: Jouve India Pvt. Ltd.
Production controller: Alhady Ali
Printed and bound in the UK using 100% Renewable Electricity at CPI Group (UK) Ltd

FSC www.fsc.org
MIX
Paper from responsible sources
FSC™ C007454

This book is produced from independently certified FSC™ paper to ensure responsible forest management.

For more information visit: www.harpercollins.co.uk/green

The publishers gratefully acknowledge the permission granted to reproduce the copyright material in this book. Every effort has been made to trace copyright holders and to obtain their permission for the use of copyright material. The publishers will gladly receive any information enabling them to rectify any error or omission at the first opportunity.

Digital downloads available at:

www.collins.co.uk/HertsforLearning/downloads

Contents

Getting Started Guide

What is greater depth, and what does it look like?

Greater-depth learning is more than cognitive content; it is complex and pays attention to mathematical structure and what lies beneath a concept or procedure. A child working at greater depth is ready and able to explore mathematical concepts in more detail. Key characteristics include being curious and asking interesting mathematical questions, making connections and identifying patterns, collaborating with others effectively, communicating their ideas with increasing clarity, being creative with their responses, and considering arguments and providing counter-arguments with proof.

Collins Greater Depth Maths

The **Collins Greater Depth Maths** resources have been written by an experienced team of maths practitioners. Features of these resources include:

- high-quality mathematical investigations that spark curiosity;
- professional development embedded to increase teacher confidence;
- ready-to-teach sequences;
- 'stumble support' to provide inclusive greater-depth opportunities for all children.

The resources are provided for three phases of the primary curriculum: Key Stage 1, Lower Key Stage 2 and Upper Key Stage 2. There are 15 tasks for each phase, all fully explained and supported.

Each task within a phase has a comprehensive set of resources. The **Teacher Guide** provides detailed guidance showing teachers how to teach and facilitate each task, with different options to suit classroom needs. Each task comprises a set of component elements and could take a full lesson or several lessons. Children are given extensive opportunities to explore the mathematical concept under investigation and to discuss and share ideas.

The innovative **Pupil Resources** contain task-specific photocopiable pages that support children at each stage of a task, to develop independence.

Upper Key Stage 2 tasks and the National Curriculum

The charts on pages 5–8 list the Upper KS2 tasks and their relevant National Curriculum objectives.

All tasks can be done in either Year 5 or Year 6, but to give some guidance, the chart suggests which Year group could focus on a given task. The allocation between Years 5 and 6 ensures good coverage of different maths topics across the Year groups. To provide teachers with some flexibility, three of the tasks are allocated to both Years 5 and 6.

In the **Teacher Guide** for each task, a small icon indicates the Year group for which the task is suggested.

Year 5 Year 6

	Year 5 National Curriculum	Year 6 National Curriculum
Number and place value		
1 **The Hidden Number** Year 5	• Read, write, order and compare numbers to at least 1 000 000 and determine the value of each digit. • Solve number problems and practical problems that involve all of the above.* * See National Curriculum statutory requirements for the full list of statements to which this refers.	• Read, write, order and compare numbers up to 10 000 000 and determine the value of each digit. • Solve number and practical problems that involve all of the above.* * See National Curriculum statutory requirements for the full list of statements to which this refers.
	Links to White Rose: Year 5 Autumn – Number: Place Value	
2 **An Ancient Number System** Year 6	• Solve number problems and practical problems that involve all of the above.* * See National Curriculum statutory requirements for the full list of statements to which this refers.	• Solve number and practical problems that involve all of the above.* * See National Curriculum statutory requirements for the full list of statements to which this refers. **Addition, subtraction, multiplication and division:** • Perform mental calculations, including with mixed operations and large numbers.
	Links to White Rose: Year 6 Autumn – Number: Place Value	
Addition and subtraction		
3 **The Magic Circle** Year 5	• Add and subtract numbers mentally with increasingly large numbers. • Solve addition and subtraction multi-step problems in contexts, deciding which operations and methods to use and why. **Multiplication and division:** • Solve problems involving addition, subtraction, multiplication and division and a combination of these, including understanding the meaning of the equals sign.	**Addition, subtraction, multiplication and division:** • Solve problems involving addition, subtraction, multiplication and division. • Use estimation to check answers to calculations and determine, in the context of a problem, an appropriate degree of accuracy.
	Links to White Rose: Year 5 Autumn – Number: Addition and Subtraction	
4 **Does It All Add Up?** Year 6	• Add and subtract numbers mentally with increasingly large numbers.	**Addition, subtraction, multiplication and division:** • Perform mental calculations, including with mixed operations and large numbers. **Algebra:** • Use simple formulae. • Express missing number problems algebraically.
	Links to White Rose: Year 6 Spring – Number: Algebra	

	Year 5 National Curriculum	Year 6 National Curriculum
Multiplication and division		
5 **Dice It Up** Year 5	• Solve problems involving addition, subtraction, multiplication and division and a combination of these, including understanding the meaning of the equals sign.	**Addition, subtraction, multiplication and division:** • Solve problems involving addition, subtraction, multiplication and division.
	Links to White Rose: Year 5 Spring – Number: Multiplication and Division	
6 **Missing Digit Multiplication** Year 5 or 6	• Solve problems involving multiplication and division, including using their knowledge of factors and multiples, squares and cubes. • Multiply numbers up to four digits by a one- or two-digit number using a formal written method, including long multiplication for two-digit numbers. • Multiply and divide numbers mentally drawing upon known facts.	**Addition, subtraction, multiplication and division:** • Solve problems involving addition, subtraction, multiplication and division. • Multiply multi-digit numbers up to four digits by a two-digit whole number using the formal written method of long multiplication.
	Links to White Rose: Year 5 Spring – Number: Multiplication and Division Year 6 Autumn – Number: Addition, Subtraction, Multiplication and Division	
7 **The Perfect Number** Year 6	• Identify multiples and factors, including finding all factor pairs of a number, and common factors of two numbers. • Know and use the vocabulary of prime numbers, prime factors and composite (non-prime) numbers. • Establish whether a number up to 100 is prime and recall prime numbers up to 19. • Solve problems involving multiplication and division including using their knowledge of factors and multiples, squares and cube.	**Addition, subtraction, multiplication and division:** • Identify common factors, common multiples and prime numbers. • *Pupils continue to use all the multiplication tables to calculate mathematical statements in order to maintain their fluency.* [non-statutory guidance]
	Links to White Rose: Year 6 Autumn – Number: Addition, Subtraction, Multiplication and Division	
Algebra		
8 **Algebra is Amazing** Year 6	**Multiplication and division:** • Solve problems involving addition, subtraction, multiplication and division and a combination of these, including understanding the meaning of the equals sign.	• Express missing number problems algebraically. **Addition, subtraction, multiplication and division:** • Use their knowledge of the order of operations to carry out calculations involving the four operations.
	Links to White Rose: Year 6 Spring – Algebra	

	Year 5 National Curriculum	Year 6 National Curriculum
Fractions		
9 **Ancient Egyptian Fractions** Year 5 or 6	• Compare and order fractions whose denominators are all multiples of the same number. • Identify, name and write equivalent fractions of a given fraction, represented visually, including tenths and hundredths.	• Use common factors to simplify fractions; use common multiples to express fractions in the same denomination. • Compare and order fractions, including fractions >1.
	Links to White Rose: Year 5 Spring – Number: Fractions Year 6 Autumn – Number: Fractions	
10 **Puzzling Percentages** Year 6	• Recognise the per cent symbol (%) and understand that per cent relates to 'number of parts per hundred', and write percentages as a fraction with denominator 100, and as a decimal. • Solve problems which require knowing percentage and decimal equivalents of $\frac{1}{2}$, $\frac{1}{4}$, $\frac{1}{5}$, $\frac{2}{5}$, $\frac{4}{5}$ and those fractions with a denominator of a multiple of 10 or 25. • *Pupils should make connections between percentages, fractions and decimals [...].* [non-statutory guidance]	• Recall and use equivalences between simple fractions, decimals and percentages, including in different contexts. **Ratio and proportion:** • Solve problems involving the calculation of percentages [for example, of measures, and such as 15% of 360] and the use of percentages for comparison.
	Links to White Rose: Year 6 Spring – Number: Percentages	
Measurement		
11 **The Red Rectangle** Year 5	• Calculate and compare the area of rectangles (including squares), and including using standard units, square centimetres (cm^2) and square metres (m^2) and estimate the area of irregular shapes. **Multiplication and division:** • Identify multiples and factors, including finding all factor pairs of a number, and common factors of two numbers. • Multiply and divide numbers mentally drawing upon known facts.	• Recognise that shapes with the same areas can have different perimeters and vice versa. **Addition, subtraction, multiplication and division:** • Perform mental calculations, including with mixed operations and large numbers. • Identify common factors, common multiples and prime numbers.
	Links to White Rose: Year 5 Autumn – Measurement: Perimeter and Area	
12 **The Tale of Two Squares** Year 6	• Calculate and compare the area of rectangles (including squares), and including using standard units, square centimetres (cm^2) and square metres (m^2) and estimate the area of irregular shapes.	• Recognise when it is possible to use formulae for area and volume of shapes. **Algebra:** • *Pupils should be introduced to the use of symbols and letters to represent variables and unknowns in mathematical situations [...].* [non-statutory guidance]
	Links to White Rose: Year 6 Spring – Measurement: Perimeter, Area and Volume	

	Year 5 National Curriculum	Year 6 National Curriculum
Geometry		
13 **Talented Tessellation** Year 5	**Geometry: position and direction:** • Identify, describe and represent the position of a shape following a reflection or translation, using the appropriate language, and know that the shape has not changed. **Measurement:** • Measure and calculate the perimeter of composite rectilinear shapes in centimetres and metres. • Calculate and compare the area of rectangles (including squares), and including using standard units, square centimetres (cm^2) and square metres (m^2) and estimate the area of irregular shapes.	**Measurement:** • Recognise when it is possible to use formulae for area and volume of shapes.
	Links to White Rose: Year 5 Summer – Geometry: Position and Direction	
14 **Battle of the Nets** Year 6	**Geometry: properties of shapes:** • Identify 3D shapes, including cubes and other cuboids, from 2D representations. • *Pupils become accurate in drawing lines with a ruler to the nearest millimetre, and measuring with a protractor. They use conventional markings for parallel lines and right angles.* [non-statutory guidance]	**Geometry: properties of shapes:** • Draw 2D shapes using given dimensions and angles. • Recognise, describe and build simple 3D shapes, including making nets. • *Pupils draw shapes and nets accurately, using measuring tools and conventional markings and labels for lines and angles.* [non-statutory guidance]
	Links to White Rose: Year 6 Summer – Geometry: Properties of Shapes	
Statistics		
15 **Distorted Data** Year 5 or 6	• Complete, read and interpret information in tables, including timetables. • *Pupils begin to decide which representations of data are most appropriate and why.*	• Interpret and construct pie charts and line graphs and use these to solve problems. • Calculate and interpret the mean as an average.
	Links to White Rose: Year 5 Autumn – Statistics Year 6 Summer – Statistics	

The components

There are three components for each task:

- **Teacher Guide**
- **Teacher's Presentation PowerPoint slides**
- **Pupil Resources**.

Teacher Guide

The **Teacher Guide** gives detailed teaching pathways for each task, including:

- guidance on key mathematical ideas;
- examples of age-appropriate children's responses;
- advice on how to model the identified strategies;
- prompts and sentence stems to scaffold children's learning and maintain engagement;
- word banks to deepen and focus vocabulary.

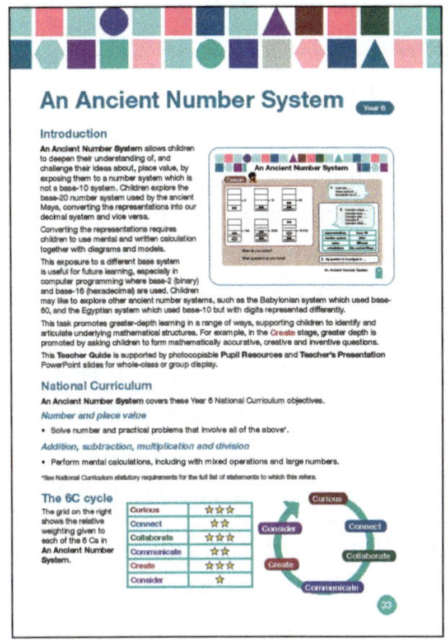

Teacher's Presentation PowerPoint slides

PowerPoint slides are provided, to enhance whole-class teaching and discussion. Some of the slides include notes providing 'stumble support' suggestions, that may help you to scaffold children in their problem-solving thinking and enable access for children who are less confident.

Where a printable handout might be useful, these are provided as black-and-white slides.

There are also slides of the **Pupil Resources**, should you wish to display these.

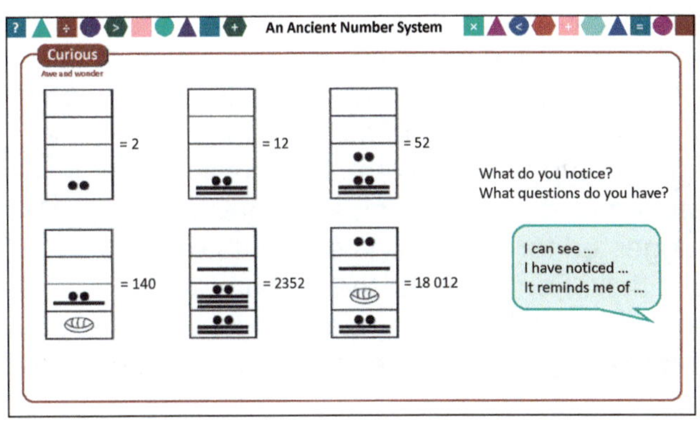

Pupil Resources

The **Pupil Resources** comprise printable pages for classroom distribution. The pages show visual representations of the problem being investigated, with prompts and word banks to help children discuss and develop their ideas. For each task there are four pages of **Pupil Resources**: one page each for the **Full** and the **Express Curious** tasks (see below for an explanation of **Full and Express Pathways**), and two pages for the remaining 5C tasks which are common to both pathways. It is suggested that teachers photocopy the pages and laminate them to create a reusable card-like resource, giving children freedom and flexibility to explore the tasks in the different Cs.

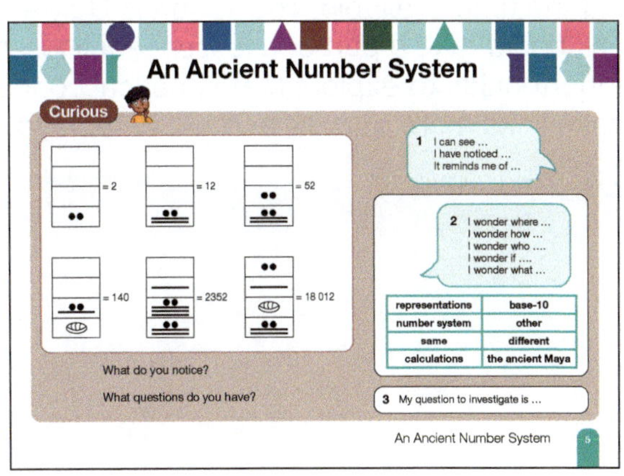

The key features of the resource

The 6C cycle

The 6Cs represent aspects of the desired characteristics of greater-depth learning: to be curious, to connect mathematical concepts, to collaborate with one another, to communicate their ideas and findings, to create their own examples and to consider and review their work and the work of others.

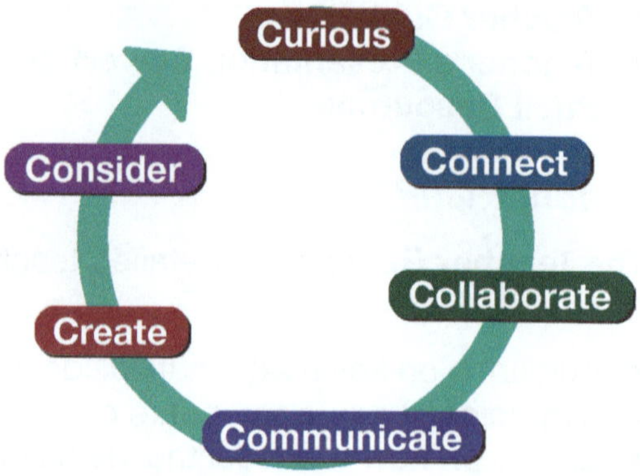

- **Curious:** promoting and encouraging curiosity in children.
- **Connect:** supporting children to connect their immediate observations to their past learning.
- **Collaborate:** developing meaningful collaboration between peers as a worthwhile skill, not just in mathematics but in everyday life.
- **Communicate:** supporting children to communicate mathematically with precision, clarity and respect.
- **Create:** fostering further curiosity and creativity.
- **Consider:** reflecting on mathematical learning experiences.

Problem-solving heuristics

Draw a diagram	✓
Make a table / spot a pattern	✓
Make a model / act it out	✓
Try then improve	
Create a systematic list	
Solve a simpler but related problem	
Work backwards	

Children are often good at applying a single strategy and repeating it, without necessarily understanding why it works, or how it might be applied in different situations. The tasks in **Greater Depth Maths** encourage the use of a range of problem-solving heuristics (strategies), and promote the discussion and investigation of the different types of strategy for different types of problem. Each task has a grid showing which strategies are suggested for the various elements within that task. For example, the grid on the right shows which strategies children are likely to use in their investigation of the task **An Ancient Number System**. These strategies are exemplified in the **Connect** stage of the task.

The route through a task

The tasks are designed to allow a degree of flexibility, according to the confidence of the children and the time available. The **Teacher Guide** offers a route through each task, using the 6Cs. Each task could take a part of a lesson, a full lesson or several lessons. Children are given extensive opportunities to explore mathematical concepts and to discuss and share ideas. Some children will make links more quickly, and will be able to skip some of the steps.

Full and Express Pathways

There are two main ways to use each task: the **Full 6C Pathway** and the **Express Pathway**. The Full Pathway exemplifies each stage of the journey through the 6Cs. However, there will be some occasions where there may not be sufficient opportunity to complete the Full Pathway, and teachers may choose to use the Express Pathway instead.

A Full Pathway could take a full lesson or several lessons, whereas an Express Pathway could form part of a lesson or could be returned to over a series of lessons.

Each task has two **Pupil Resource** pages for the **Curious** task: the first page shows the ‘Full’ version, and the second page shows a slightly modified ‘Express’ version. When following the **Full Pathway**, teachers should use the prompts in the Awe and wonder and Develop and deepen sections of the **Teacher Guide**, to encourage discussion and exploration, and allow children to spend a significant amount of time developing their own mathematical questions.

Although the **Express Pathway** prompt is similar, it is a little more direct and focused. A much shorter discussion would take place, before moving fairly quickly onto the **Connect** tasks. Teachers can then choose how much of the rest of the cycle the children engage in. Although the Express Pathway will involve fewer stages, the quality of the explanation or response should remain high. The expectation is that children would still explore the structure of the mathematics rather than just replicate the examples provided in an attempt to provide proof.

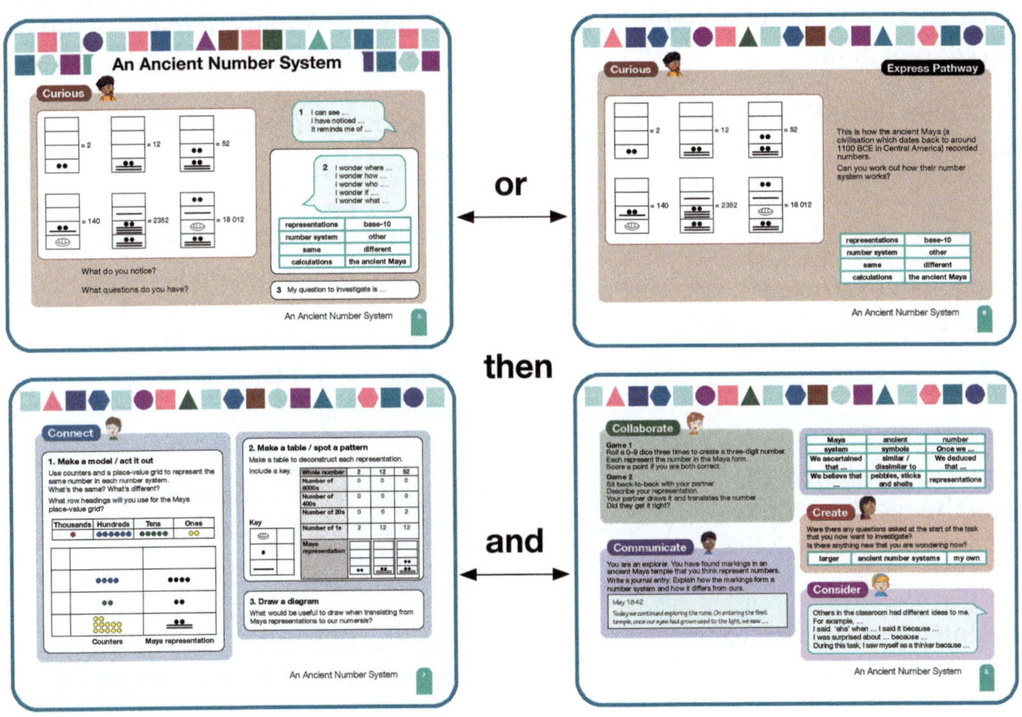

Each Pathway has its own **Curious Pupil Resource** page.

The tasks and the **Pupil Resource** pages for the remaining 5Cs are common to both pathways.

Stumble support

The in-built ‘stumble support’ provides inclusive greater-depth opportunities for all by enabling scaffold-enabled access. You will find explicit references to ‘stumble support’ within the notes section of some of the **Teacher’s Presentation** slides, but stumble support is also woven throughout all the components, in the form of word banks, prompts and sentence stems.

The **How to use** sections below explain the different 6C elements within each task, briefly describing the purpose of each element, key planning questions and details of any additional scaffolds provided. It is useful to look at the **Teacher Guide** and the **Pupil Resources** together, so that decisions concerning when to distribute the **Pupil Resources** can be made. The **Pupil Resources** do not have to be shared immediately and in some cases, the **Curious** stage will give guidance on how to introduce the question / problem to the children *before* the **Pupil Resources** are shared.

The **Teacher's Presentation** slides provide flexibility in terms of display and also ongoing 'stumble support'.

How to use the Teacher Guide

The tasks work with a series of prompts from the teacher, that can be used as whole-class prompts, or with smaller groups. These include key structural questions and prompts such as:

- *What's the same? What's different?*
- *Which is harder / which is easier?*
- What happens if I change ... ?
- If I know this, I also know ...

The prompts aim to elicit understanding of concepts, to get children thinking about how they can apply ideas to unfamiliar situations, and for teachers to assess whether children are making rich connections between different areas.

Italic bulleted text indicates questions the teacher could ask the children. →

- *How did you decide where to start?*

Other bulleted text indicates exemplar statements, questions or sentence stems from children. →

- If I built this number in base-10, I would have used one counter in the thousands column, six counters in the hundreds, five counters in the tens and two counters in the ones.

Where the first few words are **underlined**, this indicates the words the teacher could use to prompt children towards making the observation. →

- <u>I wonder</u> how this number system is the same and how it is different to ours.

Non-bulleted text indicates instructions to the teacher. →

Ask children to explain how they built up their numbers with their counters.

A **yellow warning symbol** indicates important things to watch out for: common misconceptions that need addressing in case they lead children down a false path, or things that must be done before moving on to the next part of a task. →

⚠ Watch for children who disregard each arrangement that they make instead of making an improvement each time based on what they have noticed.

The **Teacher Guide** tends to follow the same structure for each task. Below is an explanation of each section, using the task **An Ancient Number System** for illustrative purposes.

Text in blue below indicates questions teachers may want to ask themselves when they get to each section of a task.

Introduction

This explains the aims of the task and how it delivers greater depth.

An Ancient Number System

Year 6

Introduction

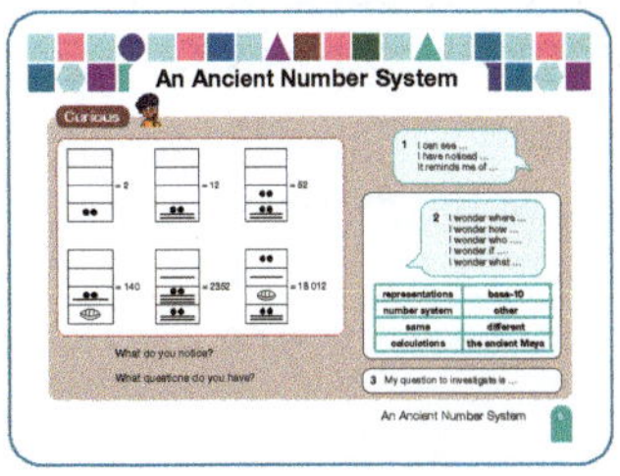

An Ancient Number System allows children to deepen their understanding of, and challenge their ideas about, place value, by exposing them to a number system which is not a base-10 system. Children explore the base-20 number system used by the ancient Maya, converting the representations into our decimal system and vice versa.

Converting the representations requires children to use mental and written calculation together with diagrams and models.

This exposure to a different base system is useful for future learning, especially in computer programming where base-2 (binary) and base-16 (hexadecimal) are used. Children may like to explore other ancient number systems, such as the Babylonian system which used base-60, and the Egyptian system which used base-10 but with digits represented differently.

This task promotes greater-depth learning in a range of ways, supporting children to identify and articulate underlying mathematical structures. For example, in the Create stage, greater depth is promoted by asking children to form mathematically accurative, creative and inventive questions.

This **Teacher Guide** is supported by photocopiable **Pupil Resources** and **Teacher's Presentation** PowerPoint slides for whole-class or group display.

National Curriculum

All of the Upper KS2 tasks can be done in either Year 5 or Year 6, but a particular Year group is suggested for each task.

Consider children's prior learning and experiences for each of the noted National Curriculum statements when deciding when to use each task.

Do children have sufficient skills and knowledge to access this task?

National Curriculum

An Ancient Number System covers these Year 6 National Curriculum objectives.

Number and place value

- Solve number and practical problems that involve all of the above*.

Addition, subtraction, multiplication and division

- Perform mental calculations, including with mixed operations and large numbers.

*See National Curriculum statutory requirements for the full list of statements to which this refers.

The 6C cycle

The relative weightings of the 6C elements are given for each task to help teachers assess which tasks are suitable for their class at different times.

The 6C cycle

The grid on the right shows the relative weighting given to each of the 6 Cs in **An Ancient Number System.**

Curious	☆☆☆
Connect	☆☆
Collaborate	☆☆☆
Communicate	☆☆
Create	☆☆☆
Consider	☆

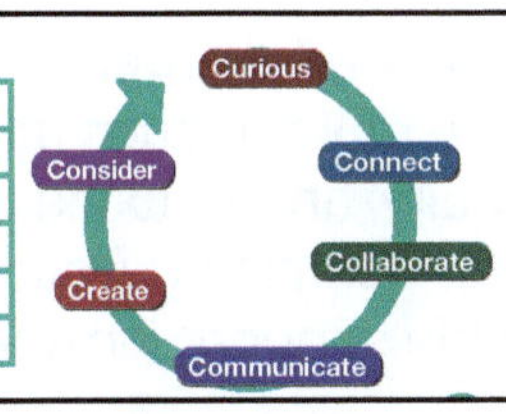

Problem-solving heuristics

For each task, specific problem-solving strategies are recommended for children to use during their investigation.

These strategies are then exemplified in the **Connect** section of the **Teacher Guide.**

Which strategies are children already aware of?

Is this an opportunity to teach a new strategy or apply a pre-existing skill?

Which strategy will be the focus of your modelling?

Problem-solving heuristics

The grid shows which problem-solving strategies children are likely to use when investigating the Connect tasks in **An Ancient Number System** These strategies are further exemplified in the Connect stage below (pages 37–39).

Draw a diagram	✓
Make a table / spot a pattern	✓
Make a model / act it out	✓
Try then improve	
Create a systematic list	
Solve a simpler but related problem	
Work backwards	

Key mathematical ideas

Notes are provided to support teachers' subject knowledge and outline the possible pedagogical pathways.

Are there any aspects you need to clarify before the teaching begins?

Key mathematical ideas

The Maya used a base-20 number system with only three symbols to represent all numbers.

These symbols are thought to represent shells, pebbles and sticks, which may have been items that were first used to count.

The Maya wrote their numbers vertically, with the highest value at the top, rather than horizontally from left to right as we do.

a shell (0)

a dot (1)

a line or bar (5)

Our base-10 place-value system uses powers of 10 (for example, ones place, tens place, hundreds place and so on), whereas in the Maya base-20 system, each base is a power of 20, for example, 1s ($20^0 = 1$), 20s (20^1 or 20×1), 400s (20^2 or 20×20), 8000s (20^3 or 20×400) and so on.

The bottom group represents numbers between 0 and 19.

The group above that represents how many 20s there are in the number.

The next group represents how many 400s there are in the number.

The next group represents how many 8000s there are in the number.

Curious

For the Full 6C Pathway, advice is provided for teachers on how to introduce the question / problem to maximise curiosity.

Two progressive elements can be used:

- **Awe and wonder;**
- **Develop and deepen.**

Awe and wonder gets children talking about the question / problem and allows them to share their initial thoughts and start to become engaged.
'Stumble support' includes:

- examples of possible responses from children, with underlined words indicating how the teacher might prompt children towards these observations.

Awe and wonder

Curious

Slide 1

Display slide 1.

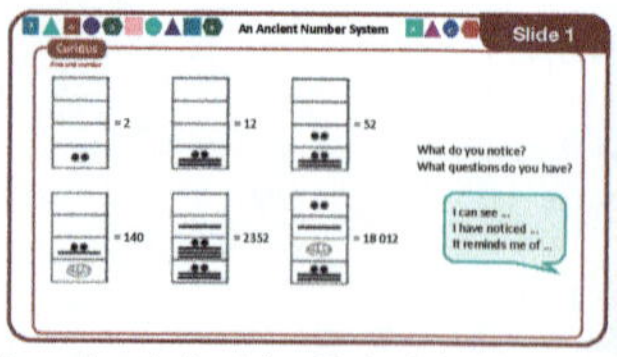

- I can see numbers and symbols.
- I have noticed that 2 is represented in each of the first 3 grids.
- I can see that some of the numbers include 0.
- I have noticed that the numbers build up vertically as they get larger, whereas we write numbers horizontally, with the digit with the greatest place value on the left.
- I have noticed that the symbols seem to be in groups.
- It reminds me of using tallies to record numbers as there are lines being used.

Develop and deepen aims to help children develop more mathematically accurate lines of enquiry, and to funnel children towards the pre-identified questions.
'Stumble support' includes:

- word banks to lift mathematical vocabulary and help direct children to appropriate lines of enquiry;
- examples of age-appropriate lines of enquiry.

How will the word bank be shared with children?

How will the possible lines of enquiry be collected? Whole class, small groups and so on. Do they need to be visible and shared with all to encourage children building on each other's ideas?

Will children all work on one line of enquiry or will they be able to select their preferred line of enquiry?

Develop and deepen

Encourage children to use the words in the word bank to develop and deepen their thoughts and questions.

number system	other
calculations	the ancient Maya
representations	base-10
same	different

Possible lines of enquiry linked to the word bank:

- I wonder where these representations come from.
- I wonder who the ancient Maya were.
- I wonder if this number system is in base-10 like ours.
- I wonder how this number system is the same and how it is different to ours.
- I wonder if we could write calculations and their answers using this number system.

Possible lines of enquiry that children may raise now, but should be held back until the Create task:

- I wonder how you would write even larger numbers.
- I wonder if we could represent all of the numbers between … and …
- I wonder if there are other ancient number systems that we could find out about.
- I wonder if I could create my own number system.

Consider whether / how to collate and display all the suggested questions, and whether / how to steer children towards selecting an appropriately challenging question.

Connect

The aim of the **Connect** task is to support children in using a range of problem-solving strategies, to try out different 'ways in' to the problem.

Guidance is given on the strategies best suited to the task and how they work in the given context.

Most **Connect** tasks will deliver a number of suitable alternative strategies. Children could work through all of the tasks suggested, or choose just one or two.

'Stumble support' includes:

- additional prompts, questions and suggestions for teachers to help children be successful with their chosen strategies.

Will children work through all of the Connect tasks, or just one or two?

What resources will need to be provided?

How will the 'stumble support' be made accessible?

Connect

Three tasks involving different strategies are suggested for children to make connections.

1. Make a model / act it out

Additional resources required: coloured counters, ThHTO place-value grid

Slide 6

Print slide 6 and distribute to children, with a ThHTO place-value grid and coloured counters.

Ask children to add their own choice of row headings (for example, '1s', '20s', '400s', '8000s', from bottom to top).

Ask children to use counters to represent the same number in both place-value grids.

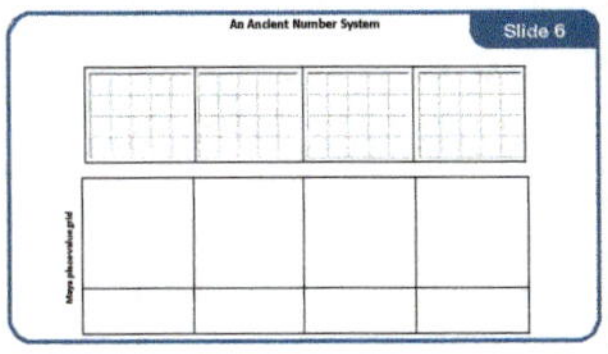

- *What's the same? What's different?*
- *In our base-10 number system we use the headings thousands, hundreds, tens and ones. The Maya system builds up vertically. What headings would you add to the rows?*
- *Think about when you regroup in the base-10 system and when you would regroup using the Maya number system.*
- There can be up to 19 counters in the ones row. Each time I reached 20, I needed to regroup the counters into the 20s row.
- There can be up to 19 counters in the 20s row. Each counter represents 20 so the highest value would be 380. If I have 20 counters, I need to regroup to the 400s row.

Collaborate

The aim of the **Collaborate** task is for children to experience working with, or sharing results with, other children.

Various models of collaboration are suggested.

Which group-work skills will be a focus for your children?

Collaborate

Refer to the Collaborate task of the **Pupil Resource**.

Additional resources required: 0–9 dice

This task provides a number of opportunities for pairs to work together to translate Maya representations into our number system or vice versa.

Slide 10

Print slide 10 and distribute to children who may need some additional practice in translating Maya numbers before playing the games in the Collaborate stage. Children can work on them independently and then check with a partner to see if they are correct.

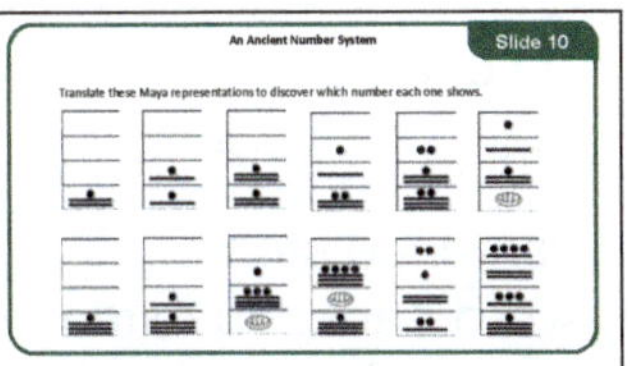

Slide 11

Display slide 11 which explains the rules for two Maya Maths games that pairs can play.

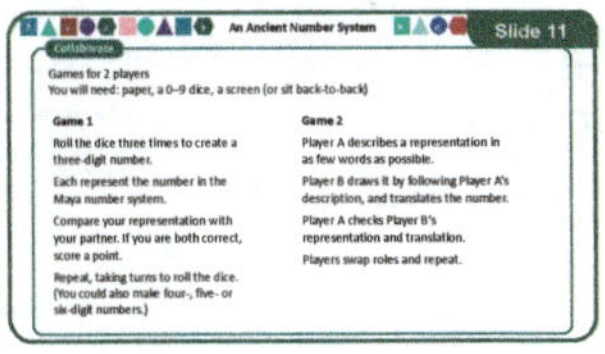

Slide 12

Print slide 12 and distribute to children. Ask them to translate the dates into Maya representations and plot them on a timeline.

Children could also work together to research explorers, as detailed in the Communicate task, or another number system such as binary or Babylonian numbers.

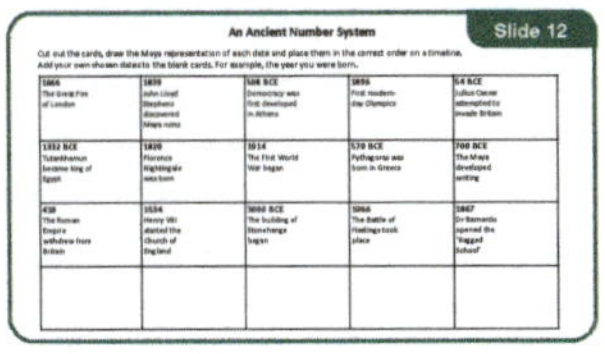

- *How did you decide where to start?*
- *Did you both start in the same way?*

Communicate

The aim of the **Communicate** task is to get children to think about what they have discovered and how they are going to share their findings with others.

There may be opportunities for cross-curricular links to the National Curriculum.

'Stumble support' varies by task but may include:

- writing frames;
- word banks to deepen understanding;
- sentence stems.

How will respectful talk be modelled?

How and where will this be recorded?

Communicate

Refer to the Communicate task of the **Pupil Resource.**

Slide 13

Display slide 13.

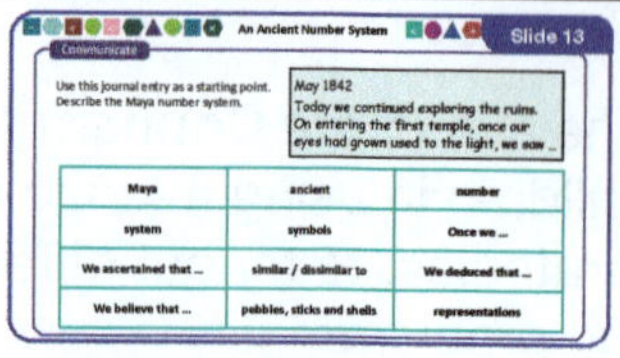

Children could write a journal entry in the role of an explorer who discovers markings inside an ancient Maya temple. Encourage children to include what they have learned about the Maya number system and the similarities and differences from their own base-10 number system. They can use the word bank for vocabulary support.

Before writing their journal entry, children may like to research John Lloyd Stephens and Frederick Catherwood, who were explorers who documented the Maya ruins in the 1840s.

Maya	ancient	number	representations
system	symbols	Once we ...	We ascertained that ...
similar / dissimilar to	We deduced that ...	We believe that ...	pebbles, sticks and shells

Create

The aim of the **Create** task is to inspire children to continue with developing their own questions to further deepen learning.

Examples of further lines of enquiry are provided in the **Teacher Guide** for each task.

Unanswered questions from the **Curious** stage can be returned to and engaged with.

'Stumble support' includes:

- sentence stems;
- word banks to extend vocabulary.

Were any additional lines of enquiry suggested in the Curious task that could be pursued now?

Create

Refer to the Create task of the **Pupil Resource.**

Slide 14

Display slide 14.

- *Were there any questions asked at the start of the task that you now want to investigate?*
- *Is there anything new that you are wondering now?*

Children can use the word bank to prompt further questions of their own.

larger	ancient number systems	my own

- I wonder how you would write even larger numbers.
- I wonder if we could represent all of the numbers between … and …
- I wonder if there are other ancient number systems that we could find out about.
- I wonder if I could create my own number system.

Slide 15

Display slide 15.

If children choose to investigate larger numbers, they will continue to explore powers of 20. So far children will have used 20^1, 20^2 and 20^3 in the tasks so far.

- *Translate this larger number.*

Support children as necessary to see that 20^4 is $20 \times 20 \times 20 \times 20$, and so the next set of symbols above would represent the number of 160 000s in the number.

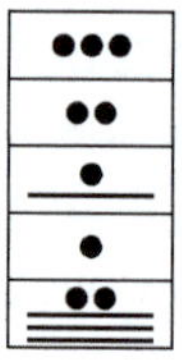

3 × 160 000 =	480 000
2 × 8000 =	16 000
6 × 400 =	2400
1 × 20 =	20
17 × 1 =	17
	498 437

Consider

The aim of the **Consider** task is to get children to reflect on what they have learned.

- Cognitive domain – what have they learned?
- Affective domain – how do they feel about the learning?
- Metacognitive domain - how did they learn it?

'Stumble support' includes:

- additional sentence stems.

Which of the suggested sentence stems will you encourage children to consider and complete?

Refer to the Consider task of the **Pupil Resource**.

Encourage children to think about the strategies that they have used. The prompts below may help get children started.

Cognitive learning domain

- Others in the classroom had different ideas to me. For example ...
- The strategies I learned during the task were ...

Affective domain

- The enjoyable bit of my learning was ... It was enjoyable because ...
- I said 'aha' when ... I said it because ...
- I was surprised about ... because ...
- I was helped by ... doing ...

Metacognitive domain

- I got stuck when ... I unstuck myself by ...
- During this task, I saw myself as a thinker because ...

Consider

Answers

Where appropriate, worked solutions and answers are provided.

For some tasks, these will be presented on the final page of the **Teacher Guide** for that task. For others, answers are given in situ throughout, because of the cumulative nature of the tasks.

Answers

Answers for slide 10

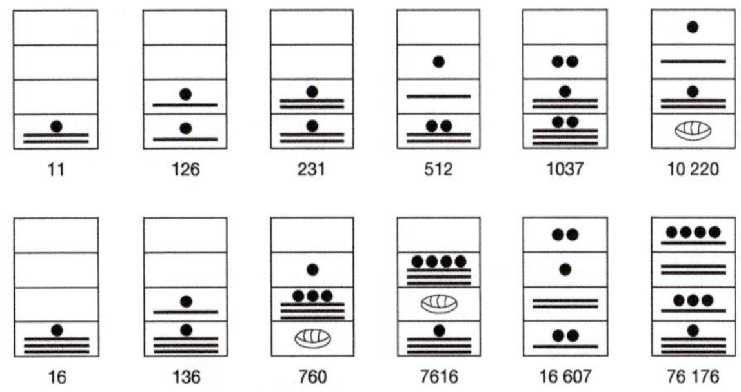

Answers for slide 12

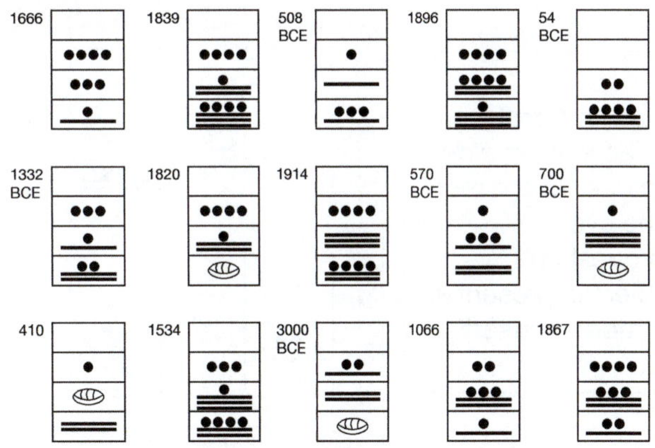

How to use the Pupil Resources

Full Pathway Curious page

Curious question / problem
This acts as an *aide memoire* for the rest of the cycle.

Awe and wonder
Sentence stems to support children to articulate their initial observations and ideas.

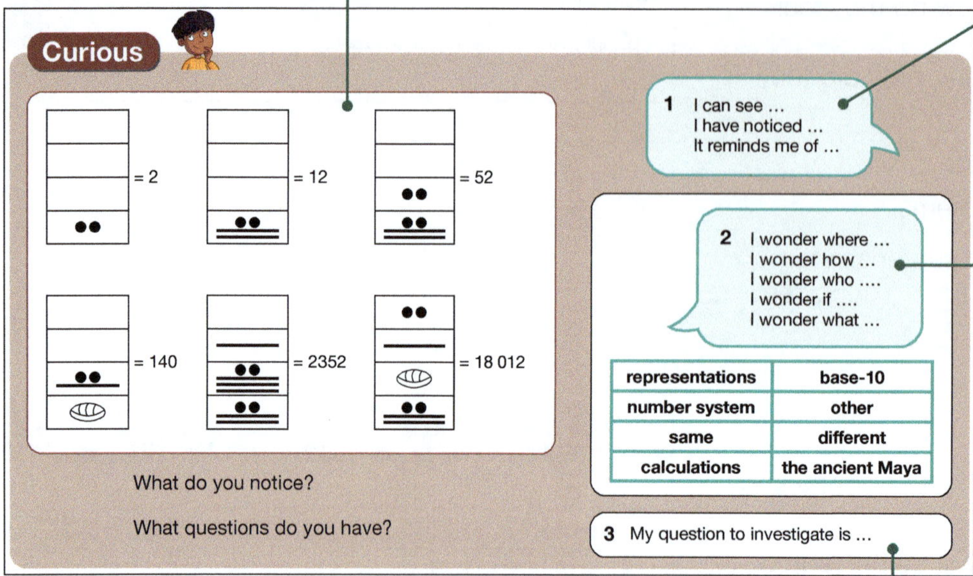

Develop and deepen
A word bank to boost mathematical vocabulary and support mathematically accurate questions. Sentence stems to support children to articulate their questions.

Chosen line of enquiry
A prompt for children to select a line of enquiry they would like to pursue.

Express Pathway Curious page

Curious question / problem
This acts as an *aide memoire* for the rest of the cycle.
The question / problem is presented in a more focused way.

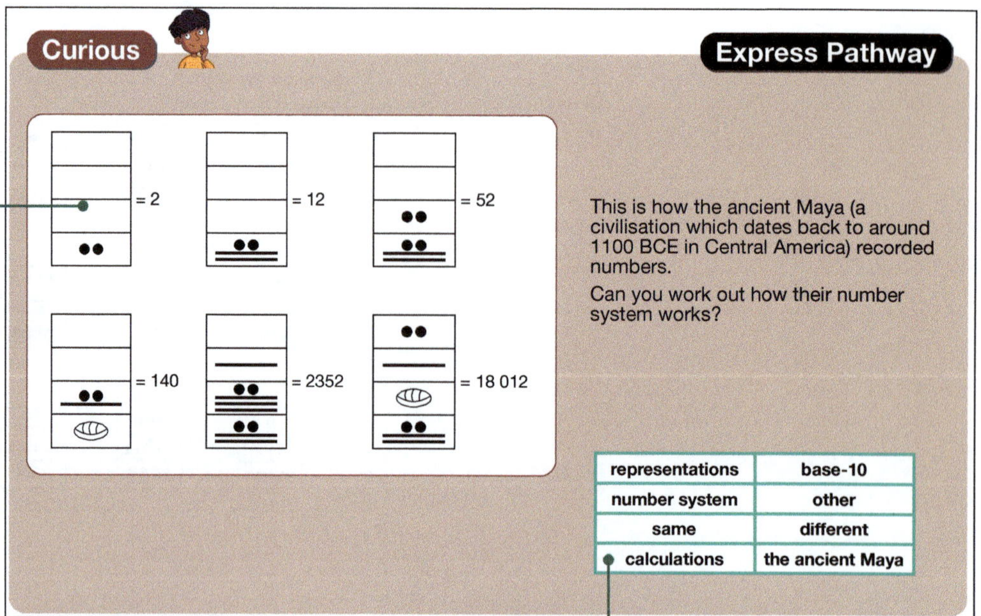

A word bank to boost mathematical vocabulary.

The remaining 5 Cs, common to both pathways

Connect
Children are presented with a range of strategies. They could select their preferred option, or try out more than one.

Connect

1. Make a model / act it out

Use counters and a place-value grid to represent the same number in each number system.
What's the same? What's different?

What row headings will you use for the Maya place-value grid?

Thousands	Hundreds	Tens	Ones
●	●●●●●●	●●●●●	○○

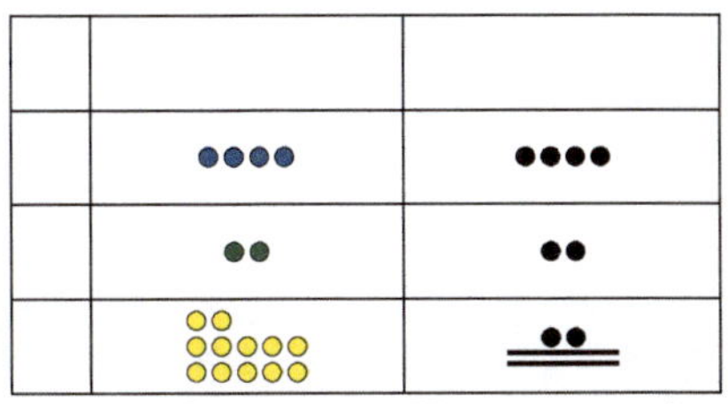

Counters Maya representation

2. Make a table / spot a pattern

Make a table to deconstruct each representation.
Include a key.

Key

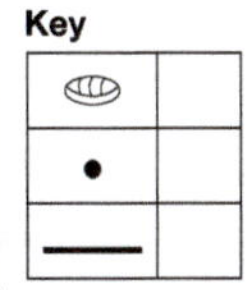

Whole number	2	12	52
Number of 8000s	0	0	0
Number of 400s	0	0	0
Number of 20s	0	0	2
Number of 1s	2	12	12
Maya representation			

3. Draw a diagram

What would be useful to draw when translating from Maya representations to our numerals?

Collaborate
Children are prompted to work with others, and consider helpful ways of working.

Create
Children are prompted to continue developing their own questions to further deepen their learning. A word bank encourages mathematically accurate language.

Collaborate

Game 1
Roll a 0–9 dice three times to create a three-digit number.
Each represent the number in the Maya form.
Score a point if you are both correct.

Game 2
Sit back-to-back with your partner.
Describe your representation.
Your partner draws it and translates the number.
Did they get it right?

Maya system	ancient symbols	number Once we …
We ascertained that …	similar / dissimilar to	We deduced that …
We believe that …	pebbles, sticks and shells	representations

Communicate

You are an explorer. You have found markings in an ancient Maya temple that you think represent numbers.
Write a journal entry. Explain how the markings form a number system and how it differs from ours.

May 1842
Today we continued exploring the ruins. On entering the first temple, once our eyes had grown used to the light, we saw …

Create

Were there any questions asked at the start of the task that you now want to investigate?
Is there anything new that you are wondering now?

larger	ancient number systems	my own

Consider

Others in the classroom had different ideas to me.
For example, …
I said 'aha' when … I said it because …
I was surprised about … because …
During this task, I saw myself as a thinker because …

An Ancient Number System 8

Communicate
Children are asked to explain their findings clearly, and are given suggestions for how their ideas could be shared. Word banks, sentence stems and writing frames are provided where appropriate.

Consider
Sentence stems guide children in reflecting on what they have learned, how they learned it and how they feel about their learning.

Assessment rubric

The assessment rubric has been designed to identify the stepping stones of attainment across the 6Cs. For a child to be described as working at greater depth, they need to be performing highly in all components. Most children working at greater depth will have a mixed profile between consolidating and extending. The rubric should be used with the whole of the class.

Using a rubric can help with:

- developing self-regulation: children can use it to help plan or monitor their own learning;
- direct teaching: it moves the focus from task completion to learning skills, knowledge and behaviour;
- providing a learning pathway: when used over time, it can help to identify gaps and next steps so feedback can be fine-tuned and highlighted as a teaching opportunity; this feedback and teaching then loops back into the next task.

Upper Key Stage 2 assessment rubric

	Curious	
	Awe and wonder	*Develop and deepen*
Not evident		
Entering	Makes limited contributions. Contributions made are not connected to the task.	Makes unconnected statements / questions when prompted.
Beginning	Makes simple contributions when prompted, e.g. recalls key fact or simple pattern.	Contributes statements rather than questions with limited mathematical content, sometimes needs prompting.
Developing	Contributes to discussion, identifies connections and patterns.	Contributes statements rather than questions, with some mathematical content.
Consolidating	Contributes to discussion. Notices and queries connections and patterns.	Offers a single or limited number of mathematical questions.
Extending		Offers a variety of good mathematical questions.

	Connect		Collaborate		
	Application	*Reasoning*	*Individual responsibility*	*Teamwork*	*Respect for others*
Not evident					
Entering	Has trouble figuring out how to solve the problem. heuristic to be chosen for them.	Not sure if their answer is correct or not. Describes what they did.	Often or always relies on others to do the work.	Only shares ideas and suggestions when prompted to.	Impolite or unkind to the group.
Beginning	Able to select an appropriate heuristic from a predefined list.	With help, can provide a correct answer. Offers an explanation but it is incomplete / unsystematic.	Participates but needs to be prompted and encouraged throughout.	Needs occasional prompting to share ideas and suggestions.	Sometimes ignores the feelings of others, e.g. interrupting, ignoring ideas.
Developing	Selects an appropriate heuristic. Can explain why they have chosen it, drawing on previous experience.	Provides a correct answer. Explains the pattern or rule and how it works. A systematic approach is present, if relevant.	Participates but needs occasional prompting. Gets upset if they don't get to do what they want to.	Shares ideas and makes suggestions but gets upset if they are rejected by the group.	Usually polite and kind to the group.
Consolidating	Selects the most appropriate heuristic and can explain choice.	Understands that there are times when more than one answer is possible. Explains the pattern or rule and how it works, using examples.	Participates within the group without arguing.	Offers suggestions and ideas. Listens carefully to other group members.	Polite and kind to group members.
Extending	Selects the most appropriate heuristic and monitors progress towards solution. If necessary, adjustments in strategy are made and / or alternative strategies are considered.	When possible, provides multiple answers. Explains the pattern or rule and how it works, using examples and counter-examples. May attempt to present a formula or algebraic proof.	Participates readily and is willing to change roles. Can identify which roles would be best suited to their strengths.	Offers ideas and suggestions, accepts if they are rejected by the group. Listens carefully and builds on other group members' views. Can identify when other team members are struggling and offers support.	Polite and kind to group members. Supports and encourages the group with energy and patience. Disagrees respectfully.

	Communicate		Create
	Clarity	*Mathematical language, images and diagrams*	
Not evident			
Entering	Communication is confusing and difficult to follow.	Attempts to use some mathematical language.	Not able to suggest any further questions.
Beginning	Communication is hard to follow, but can answer questions about learning if prompted.	Uses some mathematical language and models / diagrams.	Needs help to develop new relevant questions.
Developing	Communication is hard to follow in places.	Mathematical language is normally accurate and supported with some models / diagrams.	Builds on others' questions, changing them slightly.
Consolidating	Communication is clear and easy to follow.	Mathematical language is consistently accurate. Models / diagrams are regularly used to illustrate key ideas.	Creates their own questions.
Extending	Communication is clear, accurate and well organised. Ideas connect and flow from one to the next.	Mathematical language is consistently accurate. Models / diagrams are selected because they clarify key ideas for the audience. The flow between narrative and models / diagrams is clear and well thought out.	Creates their own questions that are creative and inventive.

	Consider		
	Cognitive	*Affective*	*Metacognition*
Not evident			
Entering	Confuses learning outcomes with title of the task.	Only states that the learning was fun or not, struggles to explain why.	Struggles to identify anything they did when they got stuck. Passively waits for an adult to support them.
Beginning	Attempts to describe what they did, but it is unclear.	Can state that the learning was fun or not; with prompting they can offer a simple reason why. Focuses on their feelings within the group.	Requires prompting to identify which learning behaviours they found helpful when they got stuck and if anyone else helped them.
Developing	Able to describe what they did.	Able to identify when things got challenging. Identifies the learning as fun or not, and can explain why. Can describe how they worked with others.	Identifies which learning behaviours they found helpful when they got stuck and, if anyone else helped them, how.
Consolidating	Able to identify what they have learned. Can see links to previous learning.	Able to explain how their feelings changed throughout the task. Describes the learning as engaging or not engaging and offer a reason why. Can describe how they tried to establish working relationships with others.	Explains which learning behaviours were useful and which were not. Includes details on what they could do differently next time.
Extending	Able to identify multiple learning outcomes. Can make explicit links between this learning and previous learning experiences, and suggest where this learning may be useful in the wider curriculum.	Reflects on how they were feeling throughout the task and why. Able to identify what was and was not engaging about the task; if appropriate they make suggestions to make the task more engaging. Able to identify how they established good working relationships with others and how their learning benefited from interactions with others.	Identifies and clarifies which learning behaviours were useful and which were not. Includes details on how they applied learning from previous tasks and what they could do differently next time.

The Hidden Number

Year 5

Introduction

The Hidden Number challenges children to identify a seven-digit number by following a complex rule.

To find the solution, they will need trial and improvement, logical thinking and perseverance. They will need to consider carefully the implications of each digit they try out. For example, writing the digit 2 above the grey 5 dictates that there must be two 5s in the hidden number. And because a 2 has been placed in the hidden number, the digit above the grey 2 must be at least 1.

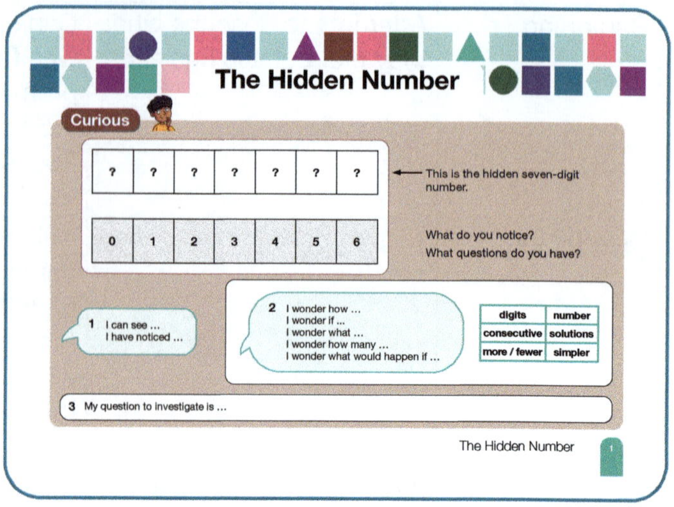

During the **Connect** stage, children could tackle simpler versions of the problem, to find hidden four-digit and five-digit numbers. They can then apply what they find out to the original seven-digit puzzle. Once this has been solved, children could extend to larger numbers. The challenge here is to discover, consider and explain patterns.

This task promotes greater-depth learning by supporting children to identify and articulate underlying mathematical structures. For example, in the **Create** stage, greater depth is promoted by asking children to form their own mathematically accurate but creative and inventive questions.

This **Teacher Guide** is supported by photocopiable **Pupil Resources** and **Teacher's Presentation** PowerPoint slides for whole-class or group display.

National Curriculum

The Hidden Number covers these Year 5 National Curriculum objectives.

Number and place value

- Read, write, order and compare numbers to at least 1 000 000 and determine the value of each digit.
- Solve number problems and practical problems that involve all of the above.*

*See National Curriculum statutory requirements for the full list of statements to which this refers.

The 6C cycle

The grid on the right shows the relative weighting given to each of the 6 Cs in **The Hidden Number.**

Curious	★★
Connect	★★
Collaborate	★★
Communicate	★
Create	★★
Consider	★★

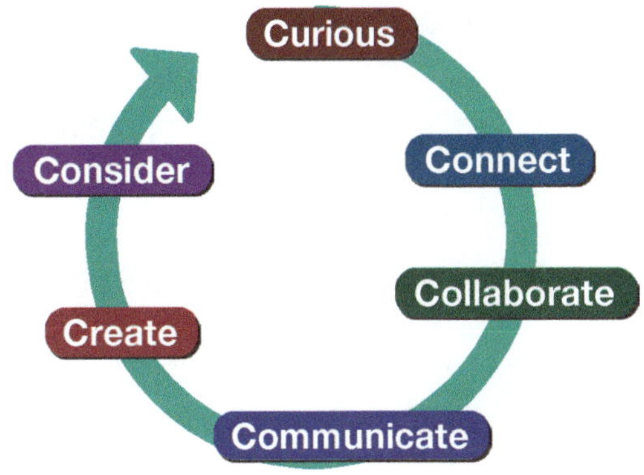

Problem-solving heuristics

The grid shows which problem-solving strategies children are likely to use when investigating the Connect tasks in **The Hidden Number.** These strategies are further exemplified in the Connect stage below (pages 28–30).

Draw a diagram	
Make a table / spot a pattern	✓
Make a model / act it out	✓
Try then improve	✓
Create a systematic list	
Solve a simpler but related problem	✓
Work backwards	

Key mathematical ideas

This task gives children an opportunity to solve a tricky problem using logical thinking. They can also rehearse reading and writing a range of numbers, many of which will include zeros as placeholders.

The hidden number is the seven-digit number indicated by question marks in the white grid. The grey grid below it shows consecutive digits from 0 to 6.

?	?	?	?	?	?	?

0	1	2	3	4	5	6

Each mystery digit must match how many times the digit in the grey cell directly below it appears in the whole number. For example, the mystery digit above the 0 must indicate the number of 0s in the whole of the hidden number.

Below is an example of what might happen when children try out a digit. In this instance, they have chosen to write '2' above the grey 5. This dictates that there must be two 5s in the hidden number, and at least one 2 (the one already written in). Wherever the 5s are written also then affects the digits below them. Let's say the child decides to put the two 5s above the grey 0 and the grey 3:

5	?	1	5	?	2	?

0	1	2	3	4	5	6

This now indicates that there are also five 0s and five 3s in the hidden number, giving 13 digits for a seven-digit number, so clearly there cannot be two 5s in the hidden number.

Here is the solution:

The hidden number contains …	three 0s	two 1s	one 2	one 3	zero 4s	zero 5s	zero 6s
	↓	↓	↓	↓	↓	↓	↓
The hidden number is **3 211 000**.	**3**	**2**	**1**	**1**	**0**	**0**	**0**
	0	**1**	**2**	**3**	**4**	**5**	**6**

Answers are given in situ throughout the tasks and are also summarised on page 32.

Teaching guidance

If you are using the **Express Pathway**, please turn to pages 28–32.

Teaching guidance for the **Full 6C Pathway** is given below.

The **Teacher's Presentation** PowerPoint file has slides specifically mentioned in this teaching guidance. Some of the slides include notes (within the PowerPoint file) providing 'stumble support' suggestions, that may help you to support children in their problem-solving thinking and enable access for children who are less confident.

Awe and wonder

Reveal slides 1–2 one at a time, allowing time for discussion and observations as prompted below.

Slide 1

Display slide 1.

Ask children to share what they are noticing or wondering.

- <u>I can see</u> one grid has digits in it and the other has question marks.
- <u>I have noticed</u> that the digits in the bottom grid are consecutive.
- <u>I can see</u> the number one hundred and twenty-three thousand, four hundred and fifty-six.

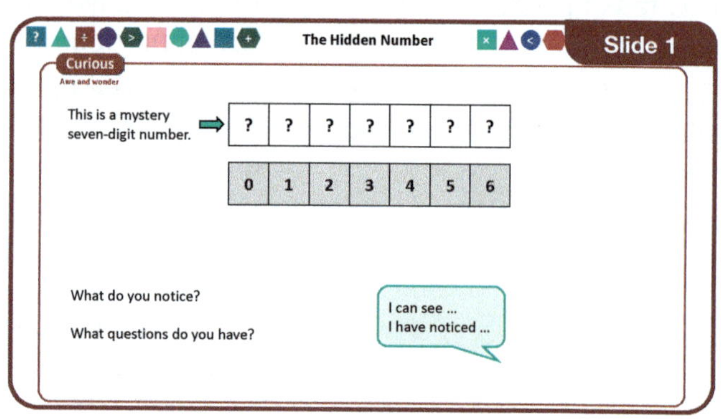

Once children have discussed their initial thoughts, explain that the question marks represent a hidden number.

Slide 2

Display slide 2 and share the rule for the hidden number:

- *To discover the hidden number, fill in the white grid so that the digit in the white box shows how many times the digit in the grey box below it occurs in the seven-digit number.*

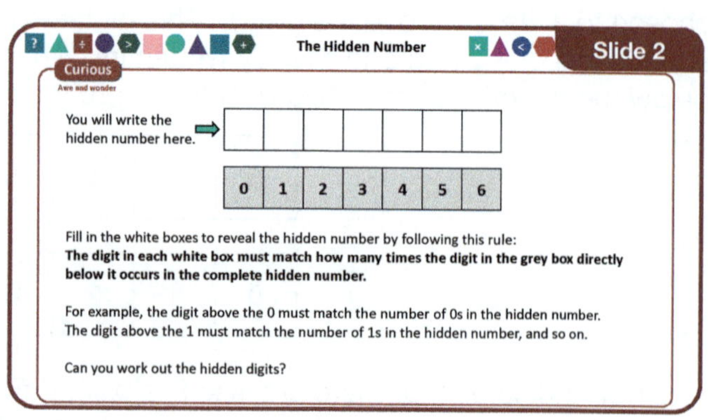

- *The number above the 0 must match the number of 0s in the hidden number. The number above the 1 must match the number of 1s in the hidden number, and so on.*

Model filling in a digit in the white grid, then discussing the implications.

For example, write a 6 above the 0.

- *What does it mean if we write a 6 above the 0?* (That the hidden number must contain six 0s.)
- *Is it possible for the hidden number to contain six 0s?*
- *The number would have to be 6 000 000, because we have already placed 6 at the start, and there are only six digits left.*
- *However, we have written the digit 6 in the number, so there is at least one 6. So we would have to write the digit 1 above the grey 6, to show that there is one 6. But we can't do that at the same time as having six 0s. So 6 cannot be an option for the first digit.*

Repeat with other digits, until all the children understand the rule.

Develop and deepen

Encourage children to use the word bank to develop and deepen their thoughts and questions.

digits	**number**	**consecutive**
solutions	**more / fewer**	**simpler**

Possible lines of enquiry:

- I wonder how you read that number.
- I wonder what the hidden number is.

Possible lines of enquiry linked to the word bank:

- I wonder if you can use all of the digits.
- I wonder how many solutions there are.
- I wonder if the digits in the hidden number are also consecutive.
- I wonder if it might be easier to solve if we try a number with fewer digits first.

Possible lines of enquiry that children may raise now, but should be held back until the **Create** task:

- I wonder if I could find a hidden number like this with 8, 9 or 10 digits.
- I wonder what would happen if we changed the order of the digits.

Consider whether / how to collate and display all the suggested questions, and whether / how to steer children towards selecting an appropriately challenging question.

Pupil Resources for Full and Express Pathways

Distribute the **Pupil Resources** for **The Hidden Number** to the children to help support them in their exploration of the different tasks. Use either of the **Curious Pupil Resources** depending on whether you are following the **Full Pathway** or the **Express Pathway**. The **Pupil Resources** for the remaining 5Cs are common to both pathways.

To provide additional flexibility in the classroom, each of the **Pupil Resources** is also provided in the **Teacher's Presentation** PowerPoint file (see slides 3–6), so you can display these at any time if you wish.

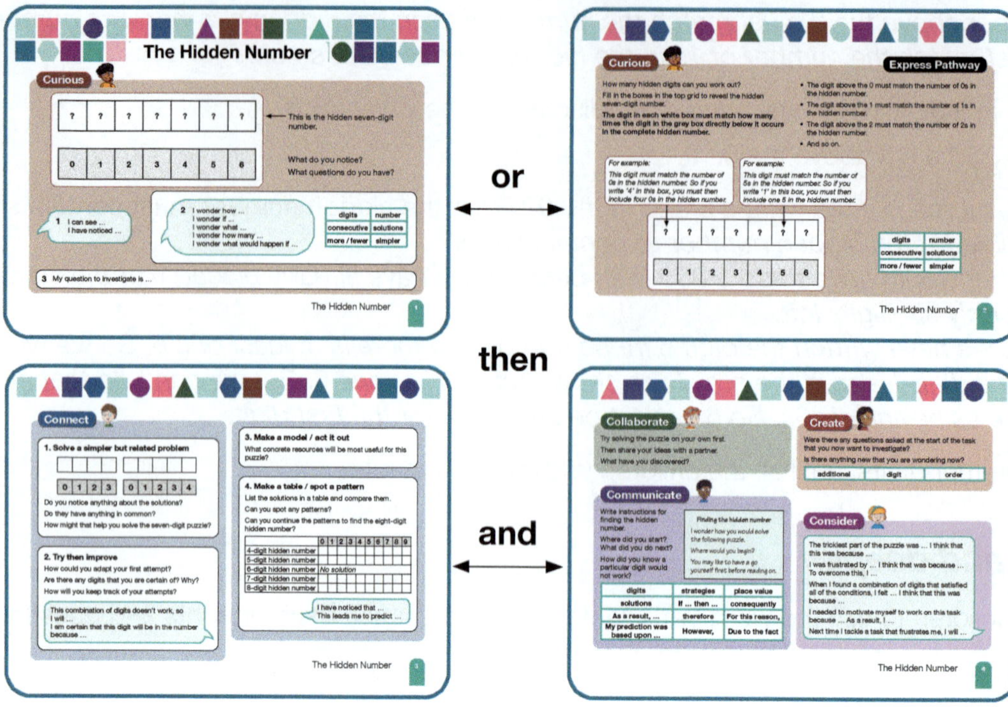

Express Pathway

If you are following the **Express Pathway**, spend a few minutes discussing the **Curious Pupil Resource** with the children. (This is slide 4, so you could display it if you wish.) Use some of the prompts on the previous page to help.

Four tasks involving different strategies are suggested for children to make connections.

1. Solve a simpler but related problem

Slide 7

Print slide 7 and distribute. Explain that they will try some simpler but related examples first to help them to get into the puzzle.

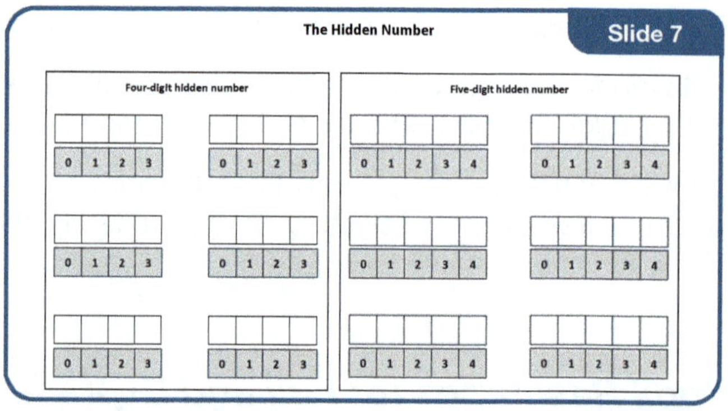

Slide 8

Display slide 8.

Ask children to try to find the four-digit hidden number. When someone thinks they have found it, ask them to explain how it fits the rule, and give the other children a chance to challenge them.

- *Can you convince us that this is correct?*
- *Does anyone else disagree?*

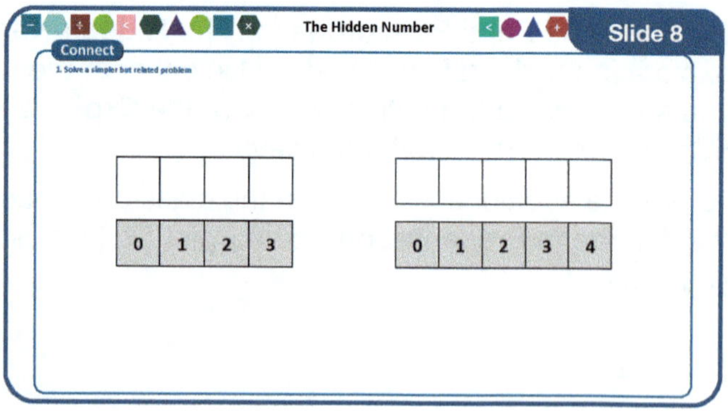

Once the correct number (1210) has been found and agreed upon, write this in the first grid on the slide.

- *Could this help us to find a solution for the five-digit puzzle?*

Once the five-digit hidden number has been identified (21 200), write this in the grid.

- *Do you notice anything about these two solutions? What is the same / different?*
- *How might that help you to solve the original puzzle?*

Slide 9

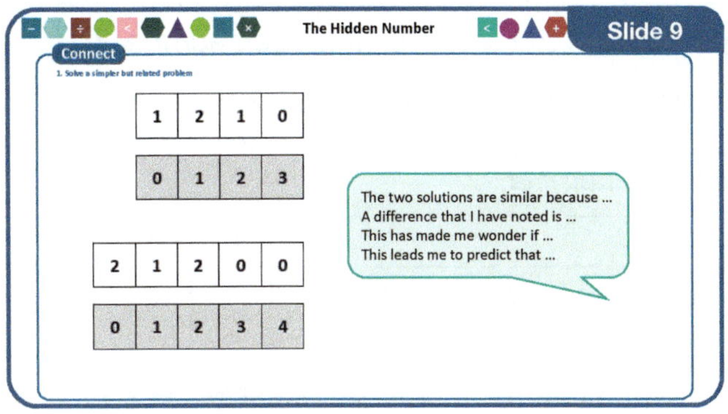

Display slide 9. Encourage children to use the sentence stems to make comparisons and predictions.

- They both end in a zero. The four-digit number has one 0 at the end, and the five-digit number has two 0s at the end. I wonder if the seven-digit puzzle will end in more than one 0.
- The highest digit used in either number is a 2. In the four-digit number, there is one 2 and two 1s. In the five-digit number, it is the other way around: two 2s and one 1. I wonder if the seven-digit puzzle will contain only 1s, 2s and 0s.

If time allows, you could give children a bit of time to explore a six-digit puzzle, but then share the fact that there is no solution to this one.

2. Try then improve

Slide 10

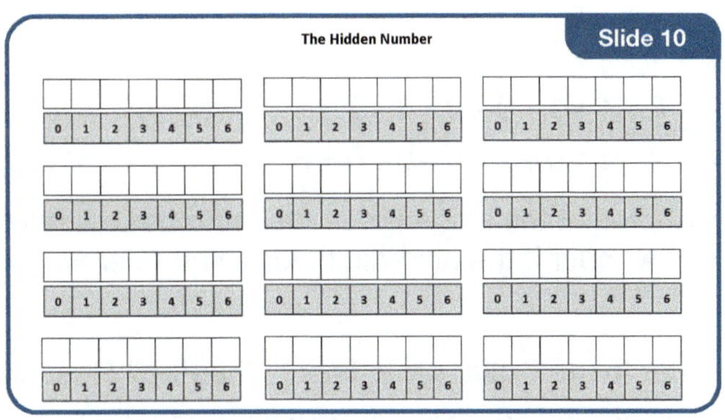

Print slide 10 and distribute to the children. Ask them to have a go at a solution using the first grid.

- *How could you adapt your first attempt?*
- *Are there any digits you know definitely can't be used? Why?*
- *Are there any digits that you are certain will be used? Why?*

Children can use the remaining grids to keep trying, aiming to improve on each attempt rather than starting from scratch.

3. Make a model / act it out

Additional resources required: digit cards 0–9

- *What concrete resources could you use to help you with this puzzle?*

Agree that digit cards are easy to move around to test out different possibilities.

Connect

4. Make a table / spot a pattern

Slide 11

Print slide 11 and distribute to the children. Ask them to fill in the solutions for the four-digit and five-digit hidden numbers (1210 and 21 200).

- *Can you use what you notice to predict the digits in the seven-digit hidden number?*

Some children may choose to continue the patterns to find a solution for an eight-digit hidden number.

The Hidden Number

Slide 11

	0	1	2	3	4	5	6	7	8	9
Four-digit hidden number										
Five-digit hidden number										
Six-digit hidden number	*No solution*									
Seven-digit hidden number										
Eight-digit hidden number										

	0	1	2	3	4	5	6	7	8	9
Four-digit hidden number	1	2	1	0						
Five-digit hidden number	2	1	2	0	0					
Six-digit hidden number	*No solution*									
Seven-digit hidden number	3	2	1	1	0	0	0			
Eight-digit hidden number	4	2	1	0	1	0	0	0		

As before, sentence stems could be used to support children to articulate any patterns they notice and make predictions as the number of digits increase.

Ask children to add up the digits in each number and compare them. They may identify the pattern in the sum of the digits. (It is equal to the number of digits in each number.)

$1 + 2 + 1 + 0 = 4$

$2 + 1 + 2 + 0 + 0 = 5$

$3 + 2 + 1 + 1 + 0 + 0 + 0 = 7$

$4 + 2 + 1 + 0 + 1 + 0 + 0 + 0 = 8$

- I predict the nine-digit hidden number will have a digit sum of 9.
- I wonder if the nine-digit hidden number will include 5, 2, 1, 1 and five 0s.

Collaborate

Refer to the **Collaborate** task of the **Pupil Resource**.

This task would work well with children initially making their first attempts independently before joining up with a partner to discuss what they have found out and which digits and numbers they have discounted. Following some paired exploration, partners could then join another pair to share thoughts and strategies.

Children could work together to create their explanation text as detailed in the **Communicate** task or when extending the puzzle as detailed in the **Create** task.

- *How did you decide where to start?*
- *Did you both start in the same way?*
- *Did you do anything differently?*

Refer to the **Communicate** task of the **Pupil Resource**.

Slide 12

Display slide 12.

This task provides an opportunity for children to write an explanation text about how they found the hidden number in order to help others. The **Pupil Resource** includes a sample opening section.

Ask children to use the prompts on the slide together with the word bank to create their text.

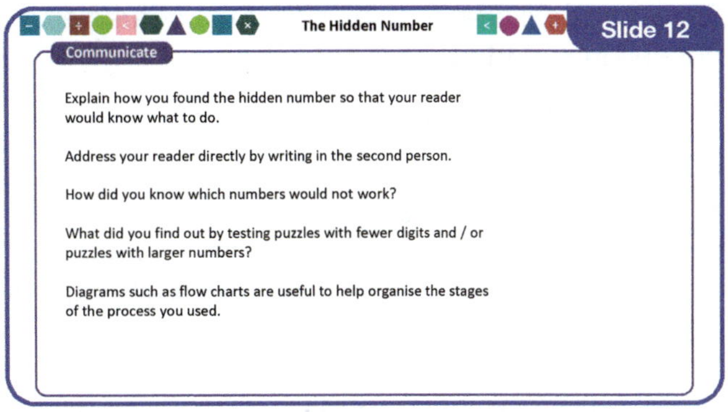

digits	**strategies**	**place value**	**solutions**
If … then …	**consequently**	**As a result,**	**therefore**
For this reason,	**My prediction was based upon …**	**However,**	**Due to the fact**

You may need to remind children about the use of the second person narrative.

Communicate

Refer to the **Create** task of the **Pupil Resource**.

Slide 13

Display slide 13.

- *Were there any questions asked at the start of the task that you now want to investigate?*
- *Is there anything new that you are wondering now?*

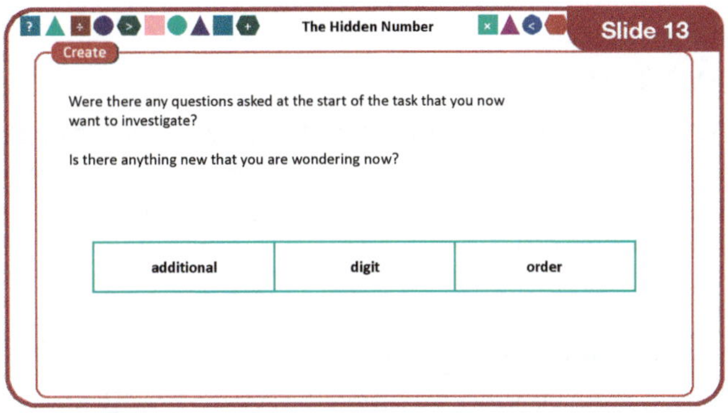

Children can use the word bank to prompt further questions of their own.

additional	**digit**	**order**

- I wonder if I could find a hidden number like this, with 8, 9 or 10 digits.
- I wonder what would happen if we changed the order of the digits.

Slide 14

Print slide 14 and distribute to children so that they can extend the puzzle if they wish.

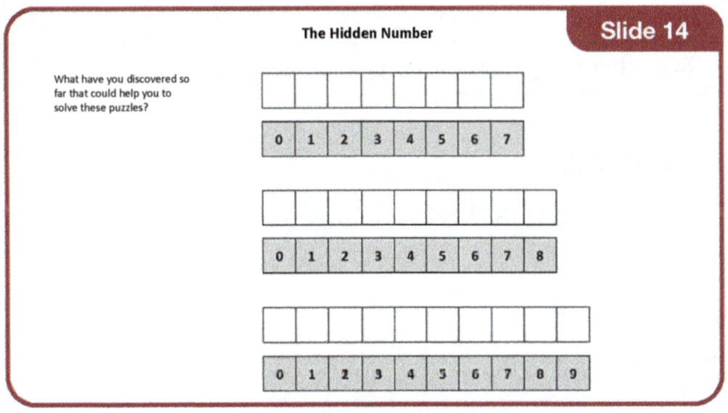

Create

Consider

Refer to the **Consider** task of the **Pupil Resource**.

Encourage children to think about the strategies that they have used. The prompts below may help get children started.

Cognitive learning domain

- The trickiest part of the puzzle was …
- I was / was not sure of my proof because …
- I was / was not sure that I had tried all possible combinations because …
- I think someone could / could not argue against my answer because …

Affective domain

- I was frustrated by … I think that was because … To overcome this, I …
- I did not give up when I got stuck. Instead, I …
- When I found a combination of digits that satisfied all of the conditions, I felt … I think that this was because …
- Next time I tackle a task that frustrates me, I will …

Metacognitive domain

- I needed to motivate myself to work on this task because … As a result, I …
- During this task, I saw myself as a thinker because …

Answers

There is only one solution to the seven-digit puzzle: 3 211 000.

	0	**1**	**2**	**3**	**4**	**5**	**6**	**7**	**8**	**9**
Two-digit hidden number	*No solution*									
Three-digit hidden number	*No solution*									
Four-digit hidden number	1	2	1	0						
Five-digit hidden number	2	1	2	0	0					
Six-digit hidden number	*No solution*									
Seven-digit hidden number	3	2	1	1	0	0	0			
Eight-digit hidden number	4	2	1	0	1	0	0	0		
Nine-digit hidden number	5	2	1	0	0	1	0	0	0	
Ten-digit hidden number	6	2	1	0	0	0	1	0	0	0

Digit sums

The digit sum is always equal to the number of digits:

$1 + 2 + 1 = 4$

$2 + 1 + 2 = 5$

$3 + 2 + 1 + 1 = 7$

$4 + 2 + 1 + 1 = 8$

$5 + 2 + 1 + 1 = 9$

$6 + 2 + 1 + 1 = 10$

An Ancient Number System

Year 6

Introduction

An Ancient Number System allows children to deepen their understanding of, and challenge their ideas about, place value, by exposing them to a number system which is not a base-10 system. Children explore the base-20 number system used by the ancient Maya, converting the representations into our decimal system and vice versa.

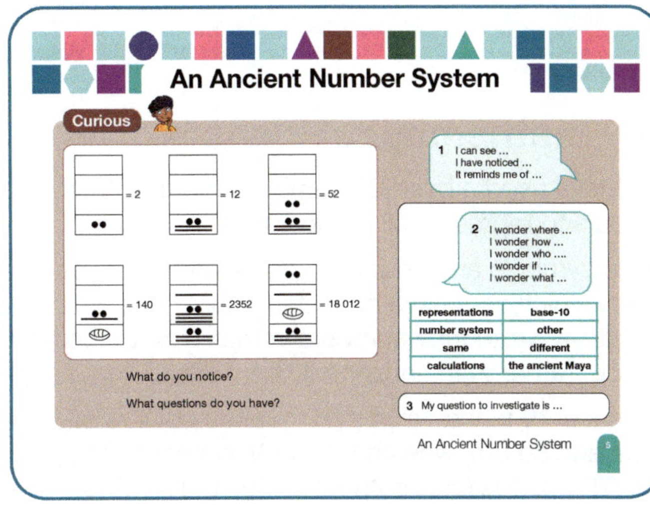

Converting the representations requires children to use mental and written calculation together with diagrams and models.

This exposure to a different base system is useful for future learning, especially in computer programming where base-2 (binary) and base-16 (hexadecimal) are used. Children may like to explore other ancient number systems, such as the Babylonian system which used base-60, and the Egyptian system which used base-10 but with digits represented differently.

This task promotes greater-depth learning in a range of ways, supporting children to identify and articulate underlying mathematical structures. For example, in the Create stage, greater depth is promoted by asking children to form mathematically accurative, creative and inventive questions.

This **Teacher Guide** is supported by photocopiable **Pupil Resources** and **Teacher's Presentation** PowerPoint slides for whole-class or group display.

National Curriculum

An Ancient Number System covers these Year 6 National Curriculum objectives.

Number and place value

- Solve number and practical problems that involve all of the above*.

Addition, subtraction, multiplication and division

- Perform mental calculations, including with mixed operations and large numbers.

*See National Curriculum statutory requirements for the full list of statements to which this refers.

The 6C cycle

The grid on the right shows the relative weighting given to each of the 6 Cs in **An Ancient Number System**.

Curious	★★★
Connect	★★
Collaborate	★★★
Communicate	★★
Create	★★★
Consider	★

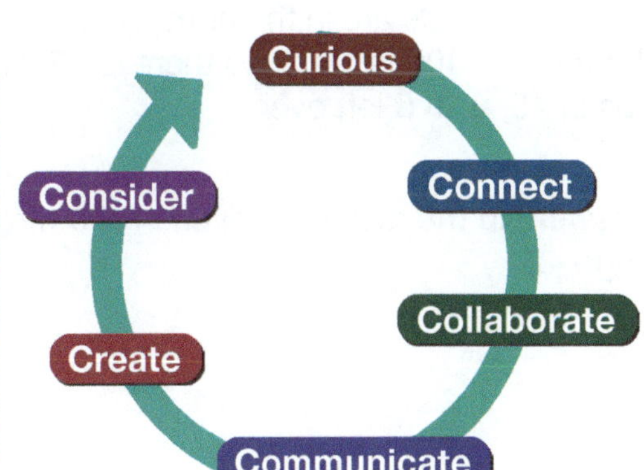

Problem-solving heuristics

The grid shows which problem-solving strategies children are likely to use when investigating the **Connect** tasks in **An Ancient Number System** These strategies are further exemplified in the **Connect** stage below (pages 37–39).

Draw a diagram	✓
Make a table / spot a pattern	✓
Make a model / act it out	✓
Try then improve	
Create a systematic list	
Solve a simpler but related problem	
Work backwards	

Key mathematical ideas

The Maya used a base-20 number system with only three symbols to represent all numbers.

These symbols are thought to represent shells, pebbles and sticks, which may have been items that were first used to count. The Maya wrote their numbers vertically, with the highest value at the top, rather than horizontally from left to right as we do.

a shell (0)

a dot (1)

a line or bar (5)

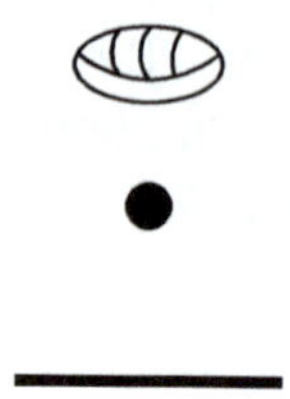

Our base-10 place-value system uses powers of 10 (for example, ones place, tens place, hundreds place and so on), whereas in the Maya base-20 system, each base is a power of 20, for example, 1s ($20^0 = 1$), 20s (20^1 or 20×1), 400s (20^2 or 20×20), 8000s (20^3 or 20×400) and so on.

The bottom group represents numbers between 0 and 19.

The group above that represents how many 20s there are in the number.

The next group represents how many 400s there are in the number.

The next group represents how many 8000s there are in the number.

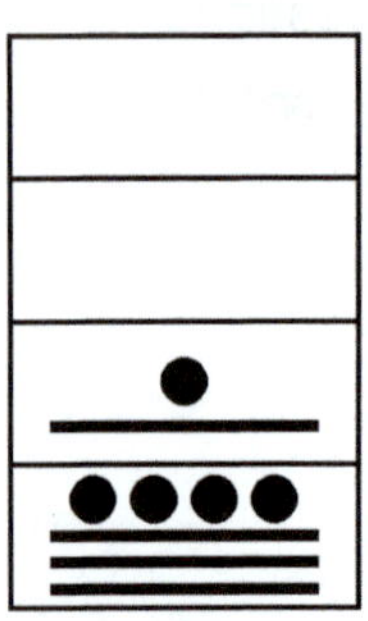

= 120 (6 × 20)

= 19 (19 × 1)

139

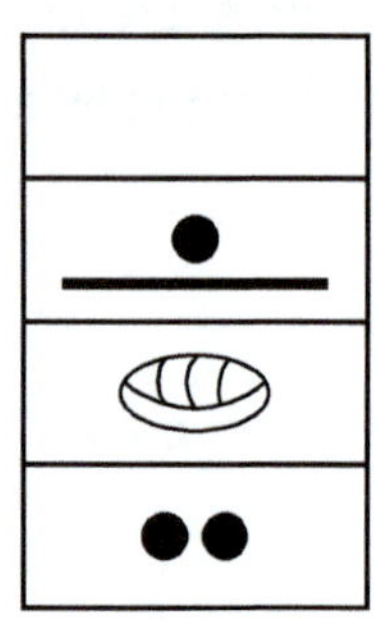

= 2400 (6 × 400)

= 0 (0 × 20)

= 2 (2 × 1)

2402

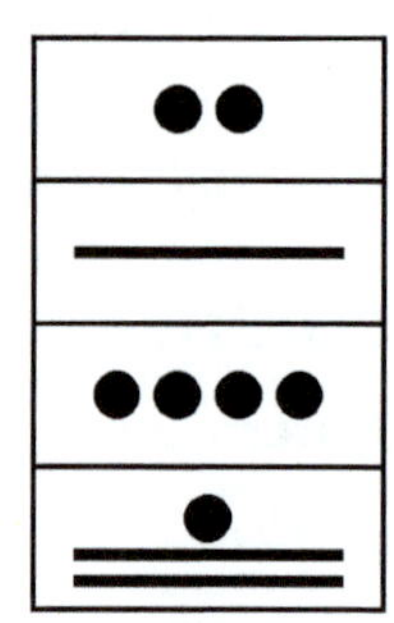

= 16 000 (2 × 8000)

= 2000 (5 × 400)

= 80 (4 × 20)

= 11 (11 × 1)

18 091

To work out the Maya equivalent of a number from our decimal system, you need to divide the number into powers of 20 (8000, 400, 20). For example, 3245: as this number is less than 8000, you'd first have to work out how many 400s are in the number (8) and show this at the top. 8 × 400 is 3200, so there is 45 left. There are two 20s in 45, with 5 left over.

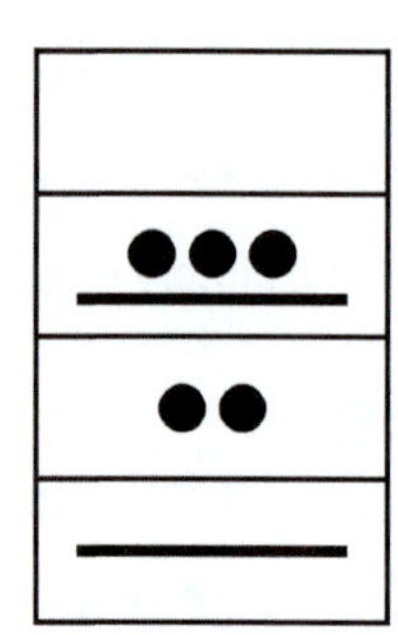

= 3200 (8 × 400)

= 40 (2 × 20)

= 5 (5 × 1)

3245

Answers to the tasks are given in situ throughout and on page 42.

Teaching guidance

If you are using the **Express Pathway**, please turn to pages 36–41.

Teaching guidance for the **Full 6C Pathway** is given below.

The **Teacher's Presentation** PowerPoint file has slides specifically mentioned in this teaching guidance. Some of the slides include notes (within the PowerPoint file) providing 'stumble support' suggestions, that may help you to support children in their problem-solving thinking and enable access for children who are less confident.

Curious

Awe and wonder

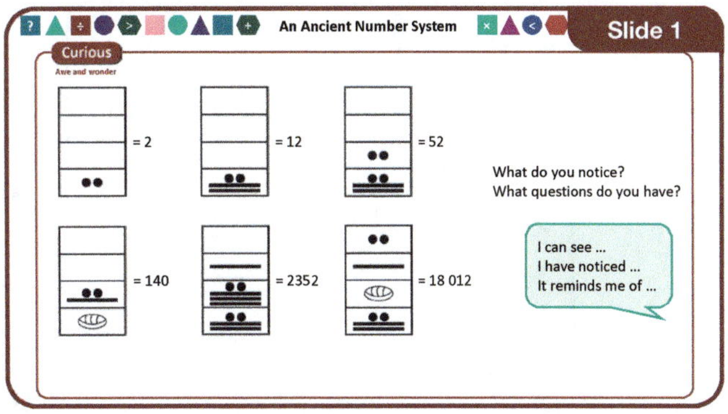

Slide 1

Display slide 1.

- I can see numbers and symbols.
- I have noticed that 2 is represented in each of the first 3 grids.
- I can see that some of the numbers include 0.
- I have noticed that the numbers build up vertically as they get larger, whereas we write numbers horizontally, with the digit with the greatest place value on the left.
- I have noticed that the symbols seem to be in groups.
- It reminds me of using tallies to record numbers as there are lines being used.

Develop and deepen

Encourage children to use the words in the word bank to develop and deepen their thoughts and questions.

number system	**other**
calculations	**the ancient Maya**
representations	**base-10**
same	**different**

Possible lines of enquiry linked to the word bank:

- I wonder where these representations come from.
- I wonder who the ancient Maya were.
- I wonder if this number system is in base-10 like ours.
- I wonder how this number system is the same and how it is different to ours.
- I wonder if we could write calculations and their answers using this number system.

Possible lines of enquiry that children may raise now, but should be held back until the **Create** task:

- I wonder how you would write even larger numbers.
- I wonder if we could represent all of the numbers between ... and ...
- I wonder if there are other ancient number systems that we could find out about.
- I wonder if I could create my own number system.

Consider whether / how to collate and display all the suggested questions, and whether / how to steer children towards selecting an appropriately challenging question.

Pupil Resources for Full and Express Pathways

Distribute the **Pupil Resources** for **An Ancient Number System** to the children to help support them in their exploration of the different tasks. Use either of the **Curious Pupil Resources** depending on whether you are following the **Full Pathway** or the **Express Pathway**. The **Pupil Resources** for the remaining 5Cs are common to both pathways.

To provide additional flexibility in the classroom, each of the **Pupil Resources** is also provided in the **Teacher's Presentation** PowerPoint file (see slides 2–5), so you can display these at any time if you wish.

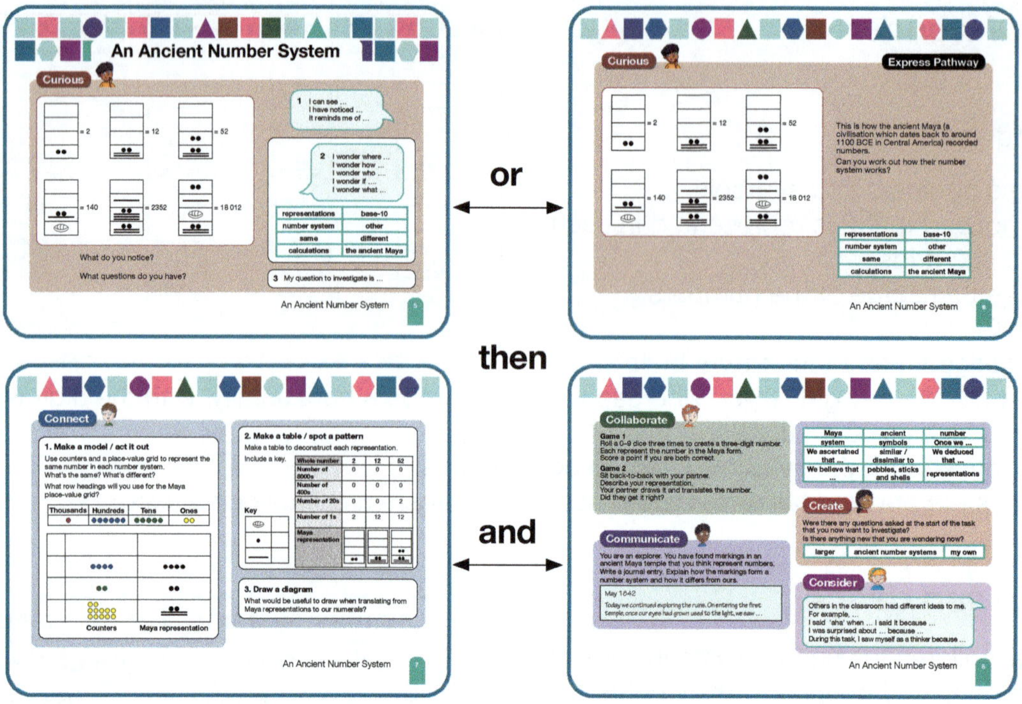

Express Pathway

If you are following the **Express Pathway**, spend a few minutes discussing the **Curious Pupil Resource** with the children. (This is slide 3, so you could display it if you wish.) Use some of the prompts on the previous page to help.

Three tasks involving different strategies are suggested for children to make connections.

1. Make a model / act it out

Additional resources required: coloured counters, ThHTO place-value grid

Slide 6

Print slide 6 and distribute to children, with a ThHTO place-value grid and coloured counters.

Ask children to add their own choice of row headings (for example, '1s', '20s', '400s', '8000s', from bottom to top).

Ask children to use counters to represent the same number in both place-value grids.

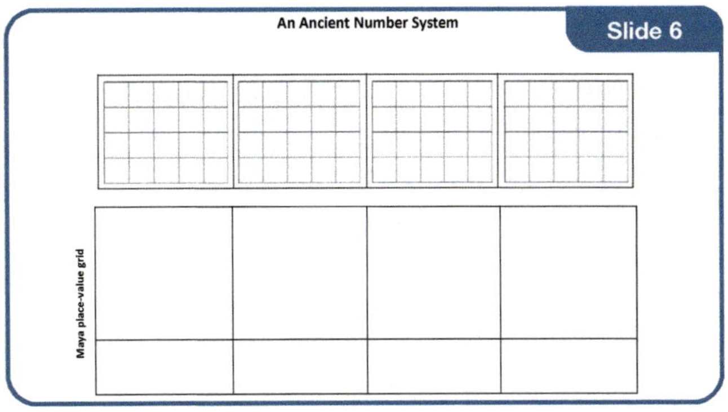

- *What's the same? What's different?*
- *In our base-10 number system we use the headings thousands, hundreds, tens and ones. The Maya system builds up vertically. What headings would you add to the rows?*
- *Think about when you regroup in the base-10 system and when you would regroup using the Maya number system.*
- There can be up to 19 counters in the ones row. Each time I reached 20, I needed to regroup the counters into the 20s row.
- There can be up to 19 counters in the 20s row. Each counter represents 20 so the highest value would be 380. If I have 20 counters, I need to regroup to the 400s row.

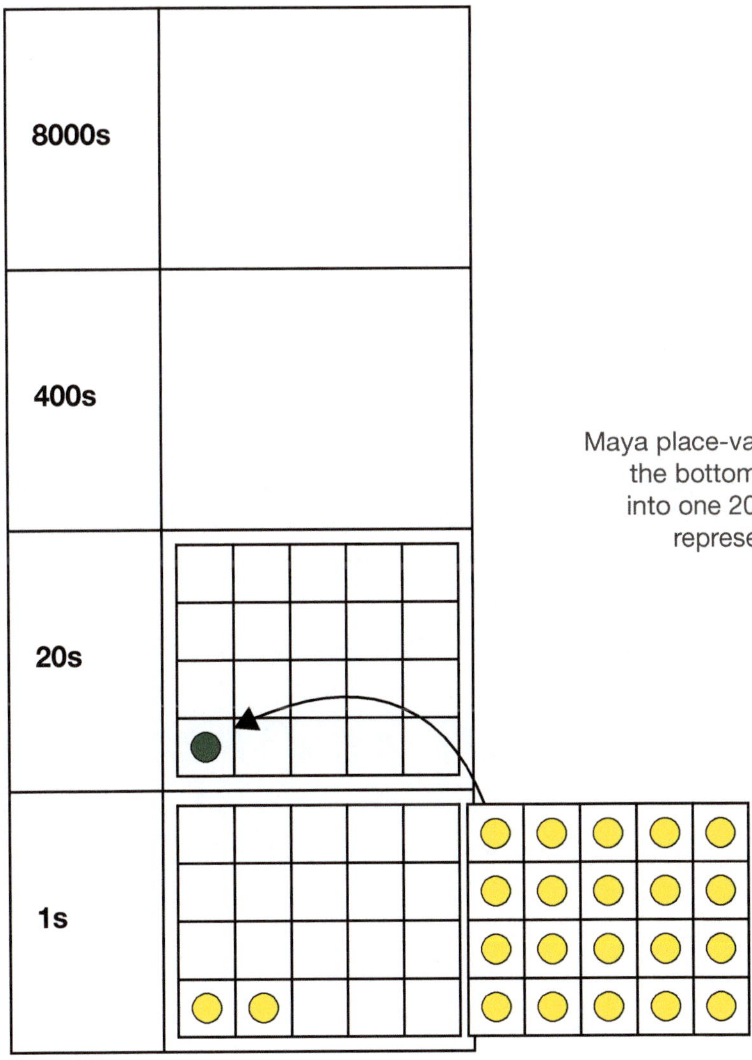

Maya place-value grid, with twenty 1s in the bottom row being regrouped into one 20 in the next row up (to represent the number 22)

Slide 7

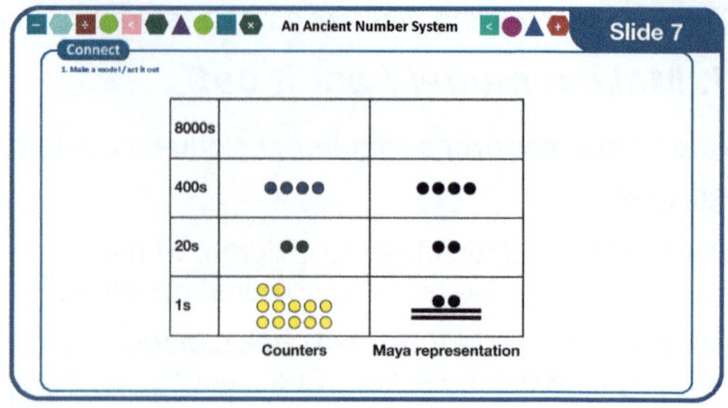

Display Slide 7.

- *What headings did you put in your place-value grid?*
- *Are they the same as these?*

Ask children to explain how they built up their numbers with their counters.

- I made 1652. I have four counters in the 400s row; 4 lots of 400 is 1600. Two counters in this row; 2 lots of 20 is 40. And there are 12 counters here; 12 ones are 12. 1600 + 40 + 12 equals 1652.
- If I built this number in base-10, I would have used one counter in the thousands column, six counters in the hundreds, five counters in the tens and two counters in the ones.

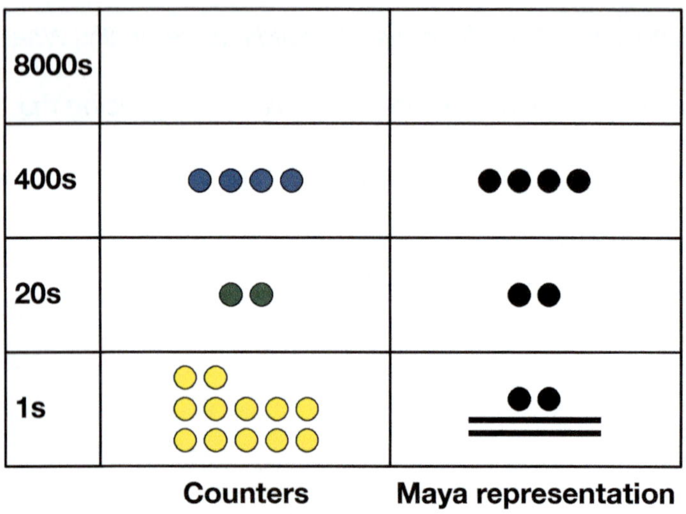

The number 1652, represented using counters in a Maya place-value grid, with the elements of the Maya representation shown alongside

2. Make a table / spot a pattern

Additional resources required: 0–9 number cards

Slide 8

Display slide 8.

Once children understand that the Maya number system is a base-20 system, a table can help to organise their thinking. Slide 8 shows one method of deconstructing each number.

Ask children to draw their own table to deconstruct all the numbers presented in the **Curious** stage. They should also complete a key to explain the symbols.

Next, ask them to generate numbers by turning over 3 or 4 number cards, fill in the table and draw the representation.

Whole number	2	12	52	140	2352
Number of 8000s	0	0	0	0	0
Number of 400s	0	0	0	0	5
Number of 20s	0	0	2	7	17
Number of 1s	2	12	12	0	12
Maya representation	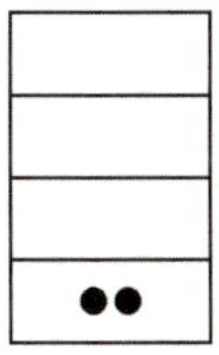	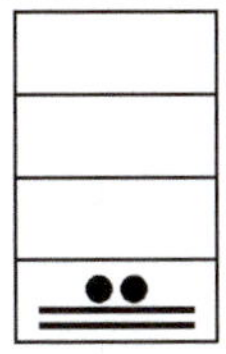	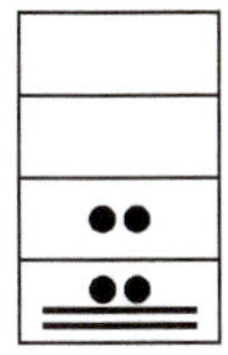		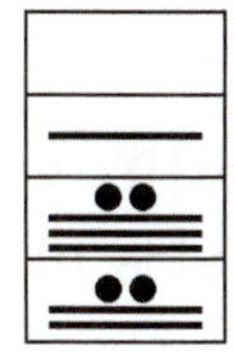

Completed slide 8 table

3. Draw a diagram

Encourage children to annotate each Maya number as they decode it to explain clearly how they know what the number represents. For example:

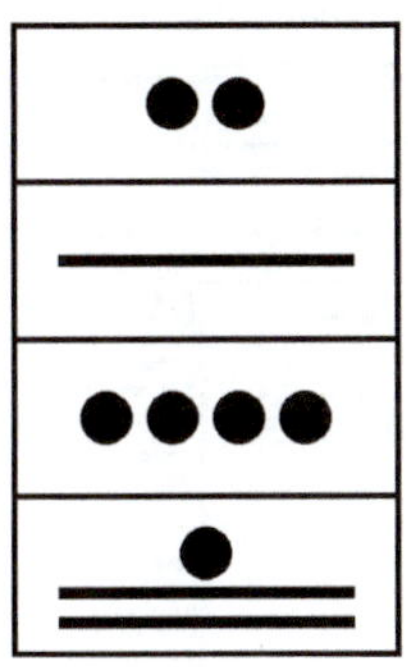

$2 \times 8000 = 16\,000$

$5 \times 400 = 2000$

$4 \times 20 = 80$

$11 \times 1 = 11$

18 091

Ask them to do the same when they convert a chosen number to the Maya representation.

Slide 9

Display slide 9.

- *Can you spot the mistakes?*

The relationship between 2000 and 5 × 400 was not considered. The correct representation would be:

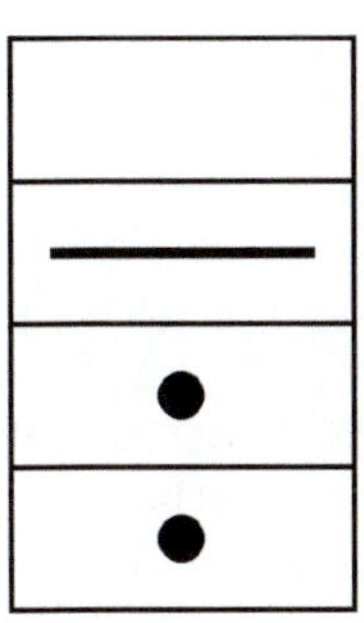

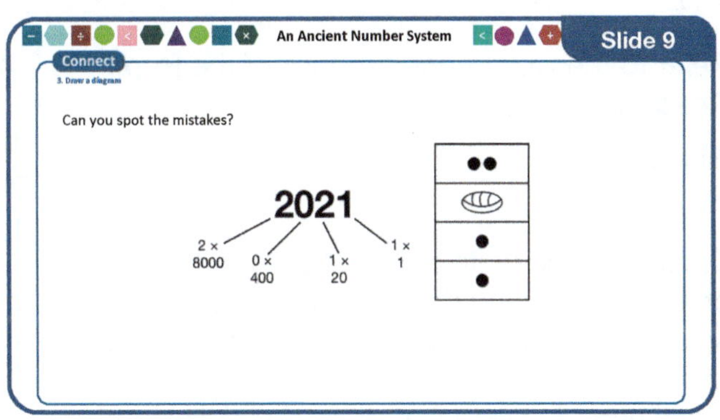

Collaborate

Refer to the **Collaborate** task of the **Pupil Resource**.

Additional resources required: 0–9 dice

This task provides a number of opportunities for pairs to work together to translate Maya representations into our number system or vice versa.

Slide 10

Print slide 10 and distribute to children who may need some additional practice in translating Maya numbers before playing the games in the **Collaborate** stage. Children can work on them independently and then check with a partner to see if they are correct.

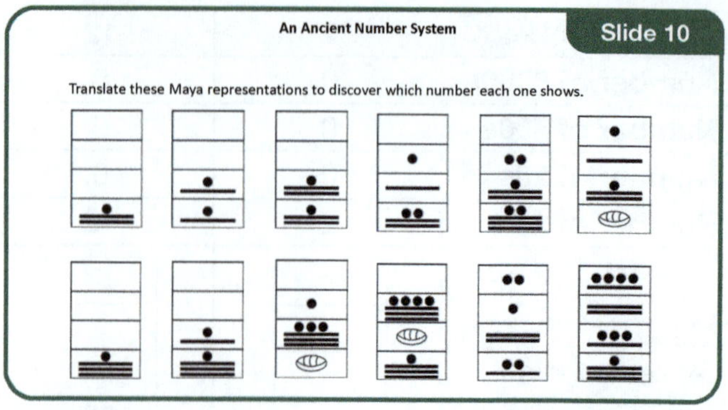

Slide 11

Display slide 11 which explains the rules for two Maya Maths games that pairs can play.

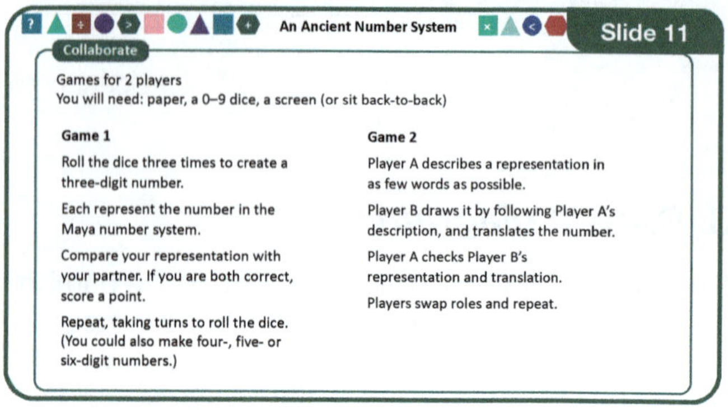

Slide 12

Print slide 12 and distribute to children. Ask them to translate the dates into Maya representations and plot them on a timeline.

Children could also work together to research explorers, as detailed in the **Communicate** task, or another number system such as binary or Babylonian numbers.

- *How did you decide where to start?*
- *Did you both start in the same way?*

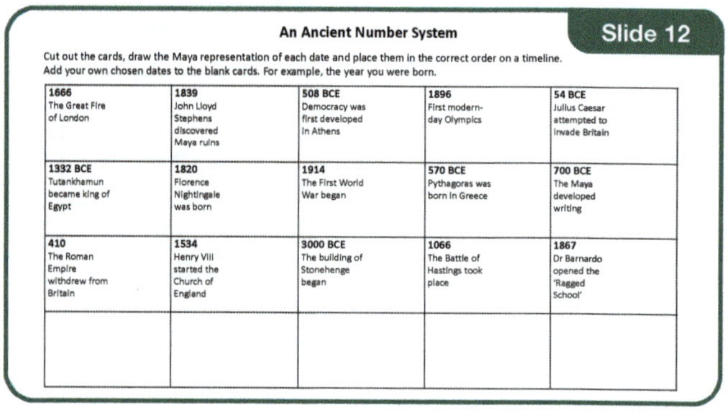

Communicate

Refer to the **Communicate** task of the **Pupil Resource**.

Slide 13

Display slide 13.

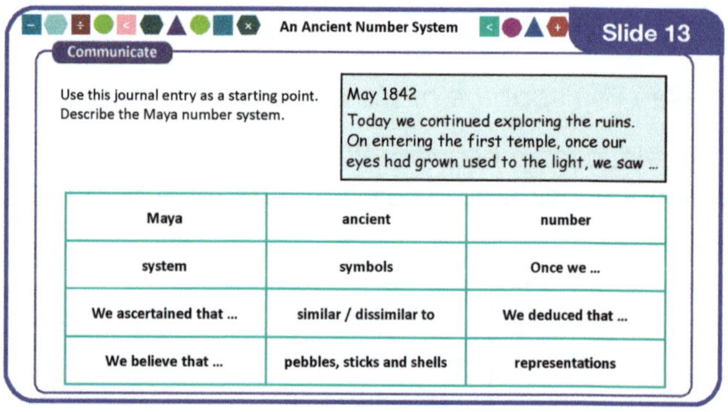

Children could write a journal entry in the role of an explorer who discovers markings inside an ancient Maya temple. Encourage children to include what they have learned about the Maya number system and the similarities and differences from their own base-10 number system. They can use the word bank for vocabulary support.

Before writing their journal entry, children may like to research John Lloyd Stephens and Frederick Catherwood, who were explorers who documented the Maya ruins in the 1840s.

Maya	**ancient**	**number**	**representations**
system	**symbols**	**Once we ...**	**We ascertained that ...**
similar / dissimilar to	**We deduced that ...**	**We believe that ...**	**pebbles, sticks and shells**

Refer to the **Create** task of the **Pupil Resource**.

Slide 14

Display slide 14.

- *Were there any questions asked at the start of the task that you now want to investigate?*
- *Is there anything new that you are wondering now?*

Children can use the word bank to prompt further questions of their own.

larger	ancient number systems	my own

- I wonder how you would write even larger numbers.
- I wonder if we could represent all of the numbers between … and …
- I wonder if there are other ancient number systems that we could find out about.
- I wonder if I could create my own number system.

Slide 15

Display slide 15.

If children choose to investigate larger numbers, they will continue to explore powers of 20. So far children will have used 20^1, 20^2 and 20^3 in the tasks so far.

- *Translate this larger number.*

Support children as necessary to see that 20^4 is $20 \times 20 \times 20 \times 20$, and so the next set of symbols above would represent the number of 160 000s in the number.

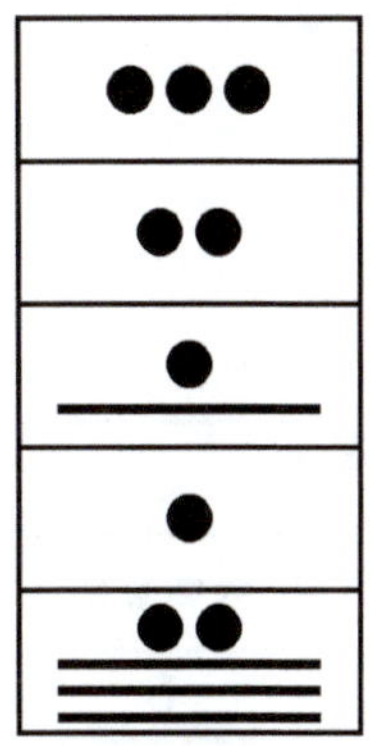

3 × 160 000 =	480 000
2 × 8000 =	16 000
6 × 400 =	2400
1 × 20 =	20
17 × 1 =	17
	498 437

Refer to the **Consider** task of the **Pupil Resource**.

Encourage children to think about the strategies that they have used. The prompts below may help get children started.

Cognitive learning domain

- Others in the classroom had different ideas to me. For example …
- The strategies I learned during the task were …

Affective domain

- The enjoyable bit of my learning was … It was enjoyable because …
- I said ‘aha’ when … I said it because …
- I was surprised about … because …
- I was helped by … doing …

Metacognitive domain

- I got stuck when … I unstuck myself by …
- During this task, I saw myself as a thinker because …

Answers

Answers for slide 10

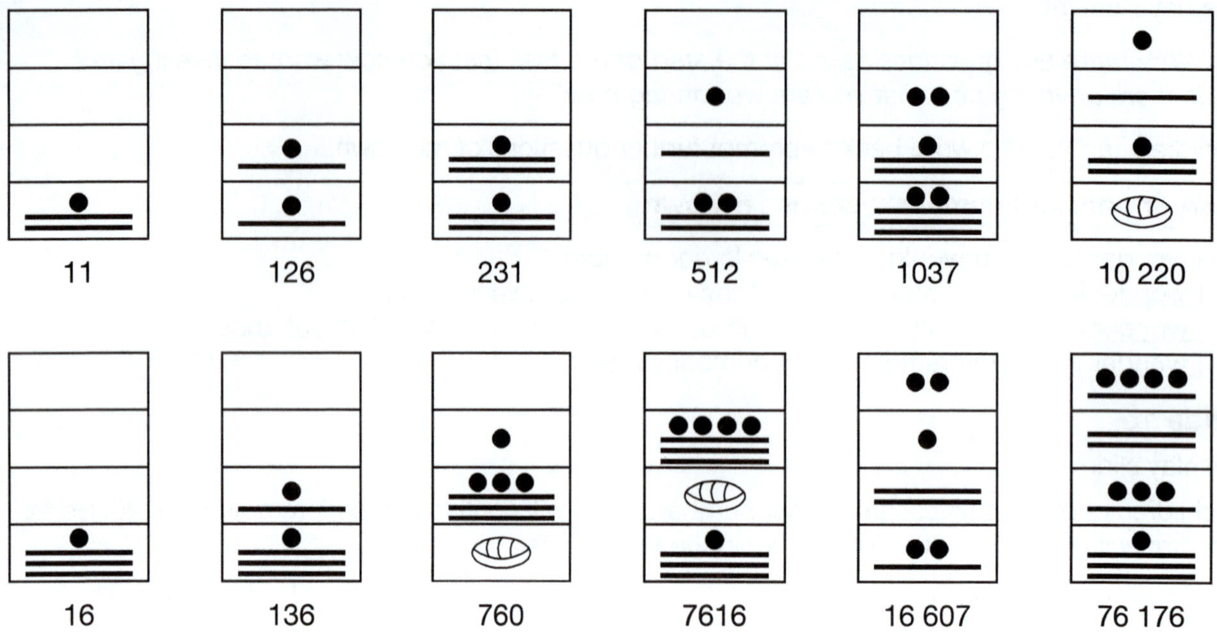

Answers for slide 12

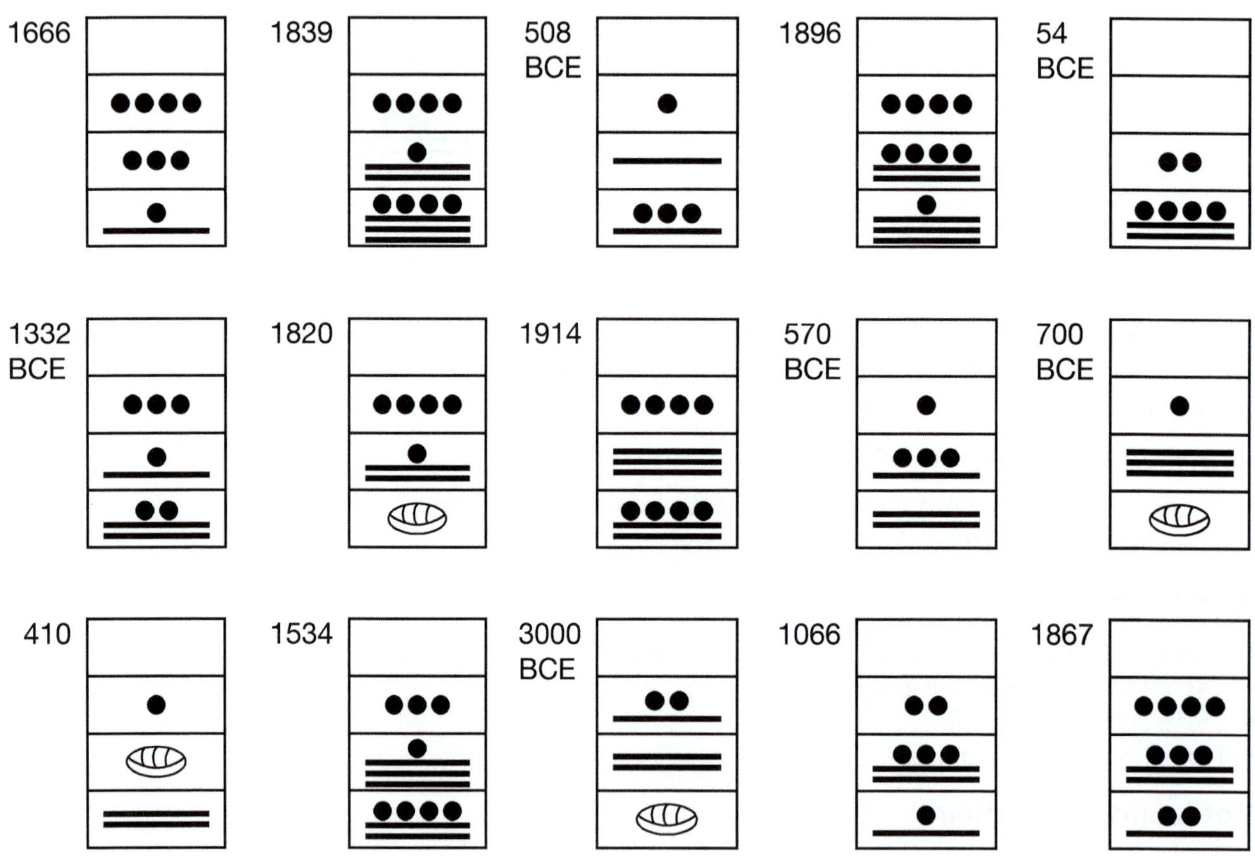

The Magic Circle

Year 5

Introduction

The Magic Circle digs deeply into number manipulation and provides an opportunity for children to rehearse and increase their fluency in addition of sets of numbers less than 20, applying mental and written methods as appropriate.

Children may choose from a range of different heuristics to solve the problem before considering if there are other possible answers and exploring additional puzzles. Throughout the task, they explain and compare their strategies.

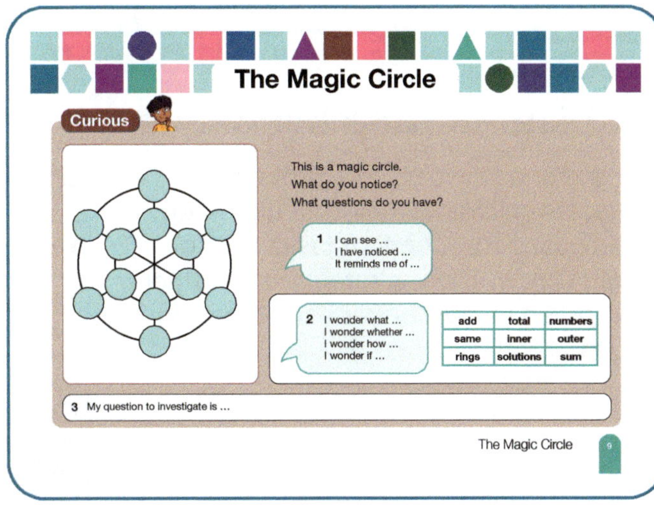

This task promotes greater-depth learning in a range of ways, supporting children to identify and articulate underlying mathematical structures. For example, in the **Connect** stage, greater depth is promoted by asking children to explore a range of different mathematical resources and models, choose appropriate resources for the task and explain their choice. They will also monitor their progress towards the solution, and adjust their strategy if needed.

This **Teacher Guide** is supported by photocopiable **Pupil Resources** and **Teacher's Presentation** PowerPoint slides for whole-class or group display.

National Curriculum

The Magic Circle covers these Year 5 National Curriculum objectives.

Addition and subtraction

- Add and subtract numbers mentally with increasingly large numbers.
- Solve addition and subtraction multi-step problems in contexts, deciding which operations and methods to use and why.

Multiplication and division

- Solve problems involving addition, subtraction, multiplication and division and a combination of these, including understanding the meaning of the equals sign.

The 6C cycle

The grid on the right shows the relative weighting given to each of the 6 Cs in **The Magic Circle**.

Curious	★★★
Connect	★★★
Collaborate	★★
Communicate	★★
Create	★
Consider	★★

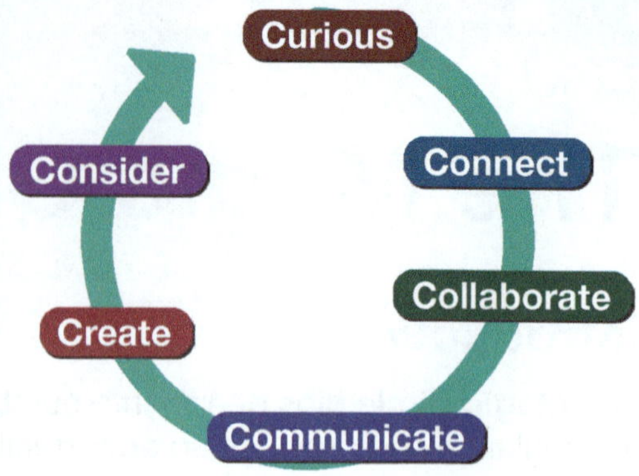

Problem-solving heuristics

The grid shows which problem-solving strategies children are likely to use when investigating the **Connect** tasks in **The Magic Circle.** These strategies are further exemplified in the **Connect** stage below (pages 47–49).

Draw a diagram	✓
Make a table / spot a pattern	
Make a model / act it out	✓
Try then improve	✓
Create a systematic list	✓
Solve a simpler but related problem	✓
Work backwards	✓

Key mathematical ideas

This task provides an opportunity for children to rehearse addition of multiple numbers under 20, applying mental and written methods as appropriate.

Children should be encouraged to consider suitable starting points to solve the puzzle. Some children may opt for a try and improve strategy as described in the **Connect** stage, trying different positioning and combinations of numbers. However, children should be encouraged to consider patterns and relationships between larger and smaller numbers and to check their calculations throughout.

To find the sum of the inner and outer rings, the total sum of the numbers should be divided by the number of rings.

$1 + 2 + 3 + 4 + 5 + 6 + 7 + 8 + 9 + 10 + 11 + 12 = 78$; $78 \div 2 = 39$; therefore each ring should sum to 39.

To find the sum of the straight lines, the total sum of the numbers should be divided by the number of lines.

$1 + 2 + 3 + 4 + 5 + 6 + 7 + 8 + 9 + 10 + 11 + 12 = 78$; $78 \div 3 = 26$; therefore each line should sum to 26.

Answers to the tasks are given on page 51.

Teaching guidance

If you are using the **Express Pathway**, please turn to pages 46–50.

Teaching guidance for the **Full 6C Pathway** is given below.

The **Teacher's Presentation** PowerPoint file has slides specifically mentioned in this teaching guidance. Some of the slides include notes (within the PowerPoint file) providing 'stumble support' suggestions, that may help you to support children in their problem-solving thinking and enable access for children who are less confident.

Awe and wonder

Slide 1

Display slide 1.

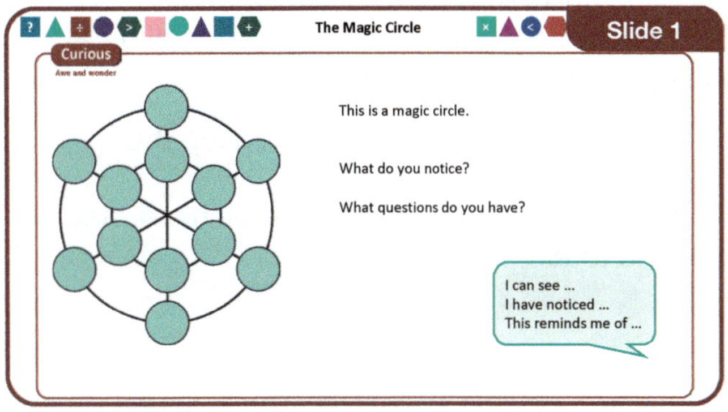

- I can see 2 circles / rings.
- I have noticed that there are 3 lines that cross in the middle.
- I have noticed that there are 6 smaller circles in the inner ring and 6 in the outer ring.
- I can see that there are 12 smaller circles in total.
- It reminds me of a magic square because they have similar names.
- It reminds me of the patterns that a spider web might make.

Develop and deepen

Encourage children to use the words in the word bank to develop and deepen their thoughts and questions.

add	**total**	**numbers**
same	**inner**	**outer**
rings	**solutions**	**sum**

Note that some children may misuse the term 'sum' to describe any calculation rather than the correct meaning of adding two or more numbers.

Possible lines of enquiry:

- I wonder what a magic circle is.

Possible lines of enquiry linked to the word bank:

- I wonder what is magic about the circle.
- I wonder whether we add numbers like with a magic square.
- I wonder whether both of the rings have the same total.
- I wonder whether the three lines sum to the same total.
- I wonder whether the three lines and the 2 rings sum to the same total.
- I wonder whether we use the numbers 1 to 12 as there are 12 circles.
- I wonder how many different solutions there are.

Possible lines of enquiry that children may raise now, but should be held back until the **Create** task:

- I wonder what would happen if we added another ring or another line.
- I wonder if it would work with the numbers 10 to 22 instead of 1 to 12.

Consider whether / how to collate and display all the suggested questions, and whether / how to steer children towards selecting an appropriately challenging question.

Slide 2

Once the children have come up with their Develop and deepen questions, display slide 2 and confirm the parameters of the task.

- The inner and outer rings both have the same total.
- The vertical and diagonal lines all have the same totals.

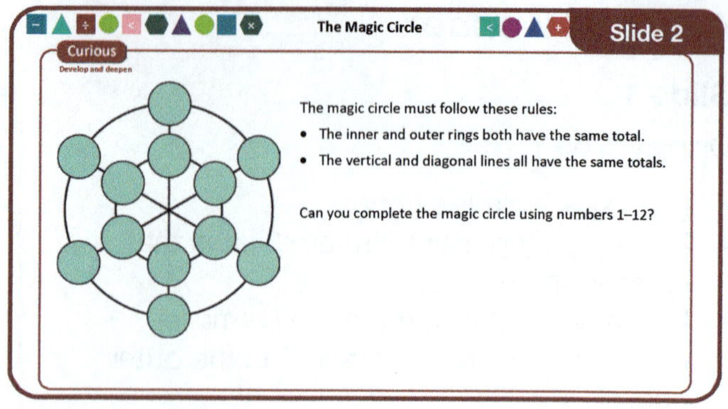

Pupil Resources for Full and Express Pathways

Distribute the **Pupil Resources** for **The Magic Circle** to the children to help support them in their exploration of the different tasks. Use either of the **Curious Pupil Resources** depending on whether you are following the **Full Pathway** or the **Express Pathway**. The **Pupil Resources** for the remaining 5Cs are common to both pathways.

To provide additional flexibility in the classroom, each of the **Pupil Resources** is also provided in the **Teacher's Presentation** PowerPoint file (see slides 3–6), so you can display these at any time if you wish.

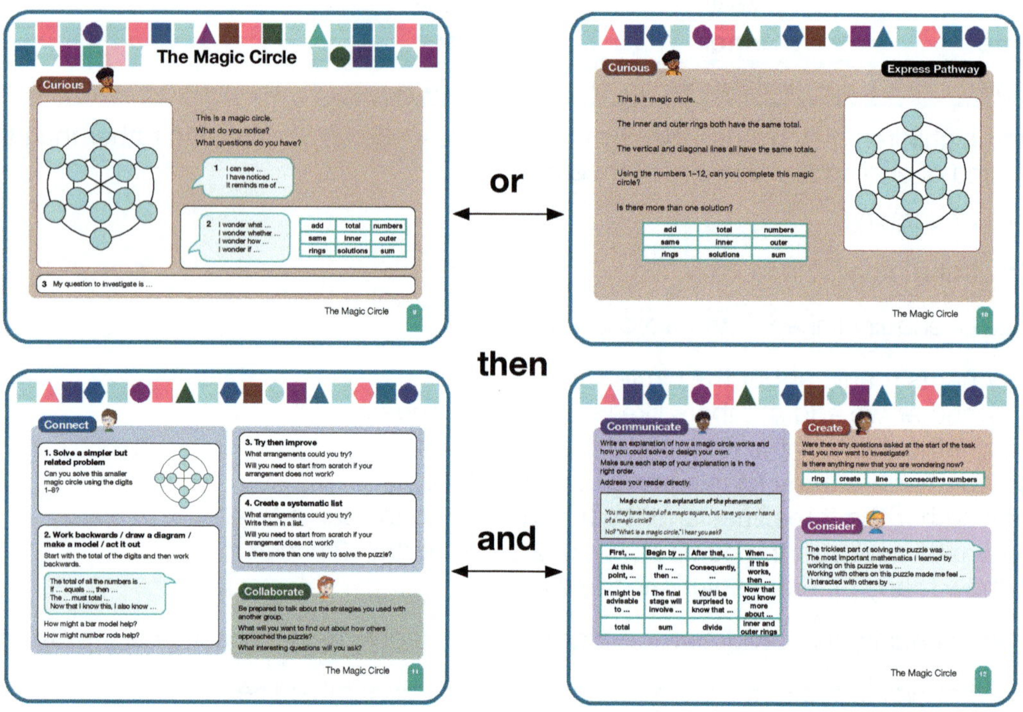

Express Pathway

If you are following the **Express Pathway**, spend a few minutes discussing the **Curious Pupil Resource** with the children. (This is slide 4, so you could display it if you wish.) Use some of the prompts on the previous page to help.

Four tasks involving different strategies are suggested for children to make connections.

1. Solve a simpler but related problem

Slide 7

Print slide 7 and distribute to children. This contains a similar but related problem that children can solve before tackling the twelve-number magic circle. The digit cards can be cut out, placed and rearranged on the magic circle.

Ask children to use the numbers 1 to 8 on this smaller version of the magic circle. Working backwards, making a model and trial and improvement could all be used as strategies, as detailed below.

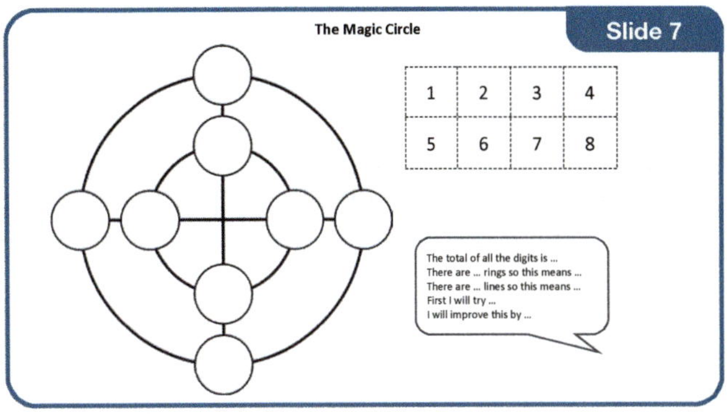

2. Work backwards / draw a diagram / make a model / act it out

Additional resources required: number rods

Slide 8

Display Slide 8.

Support children to work backwards by first finding the total of all the numbers, and then dividing this total between the number of rings and the number of lines.

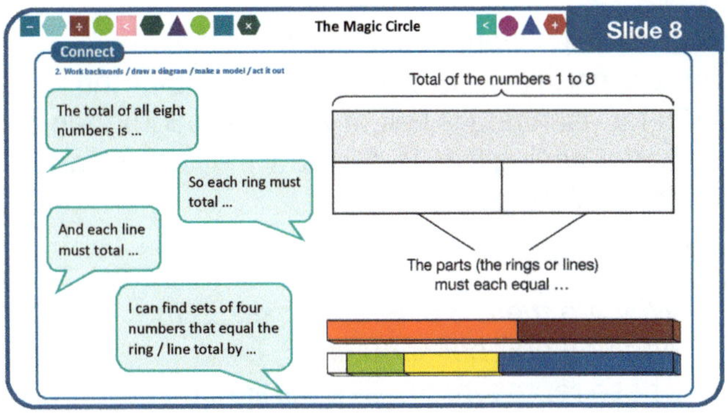

- *What happens when we add up all the numbers 1–8? What total do we get?* (36)
- *If there are 2 rings and they both need to have the same total, what does that tell us that we need to do?* (Divide 36 by 2.)
- Each ring will total 18.
- *Would a bar model help? What is the whole? What are the parts?*
- The whole is 36. The parts (the rings or lines) must be 18.

When children have ascertained that the two rings will sum to 18, and the two lines will also add up to 18, number rods could be used to explore different combinations of numbers to make the ring or line total.

- If each ring equals 18, I am going to try four numbers that add up to 18. These could go in the inner ring.
- The left-over numbers will go on the outer ring. Now I will see if each line also adds up to 18.
- *If each line totals 18, then what could half of each line total?*
- *What do we know about complements to 9?*

Slide 9

Once children have used the simpler but related prompt, the same structure could be used to explore the original prompt.

Display slide 9 which contains sentence frames to support children when working backwards.

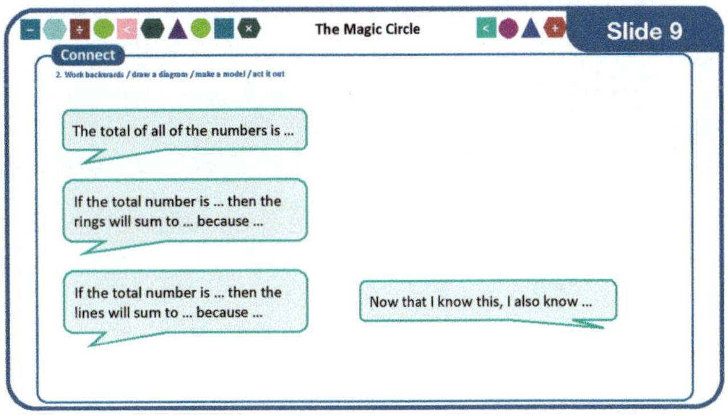

3. Try then improve

Slide 10

Print out slide 10 and distribute to children.

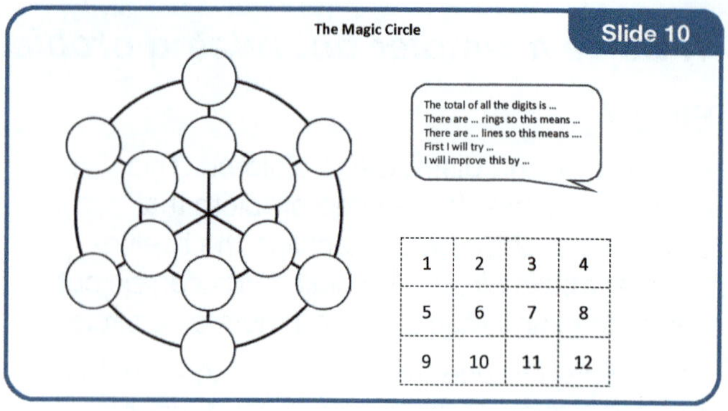

Try then improve would be a useful strategy to solve this problem and will work alongside the other examples detailed in the **Connect** stage.

Children could choose to cut out and use the digit cards, or write the numbers on sticky notes or similar, to move around into different arrangements.

Encourage children to record their attempted arrangements to avoid replications.

If using try then improve as a strategy along with another strategy, encourage children to not disregard each arrangement that they make, but instead, use it to make one improvement each time.

4. Create a systematic list

Slide 11

This strategy would be useful once children have ascertained, or been prompted, that the total for each ring is half of 78. They might then consider writing lists to see how they can total 39 for each ring.

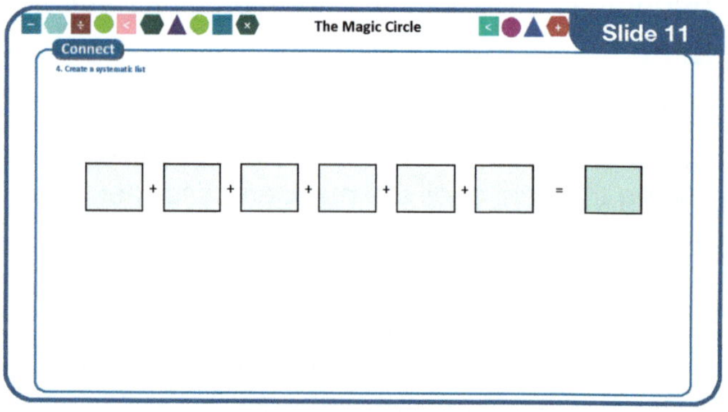

Display slide 11.

- *Why are there 6 different boxes in the addition number sentence?*
- These represent the 6 numbers on each ring.
- *What numbers could we try first in these 6 boxes?*

Slide 12

Display slide 12 and discuss the two lists of 6 addends and each total. Ask children to comment on how the numbers have been organised and whether they satisfy the rules for the magic circle.

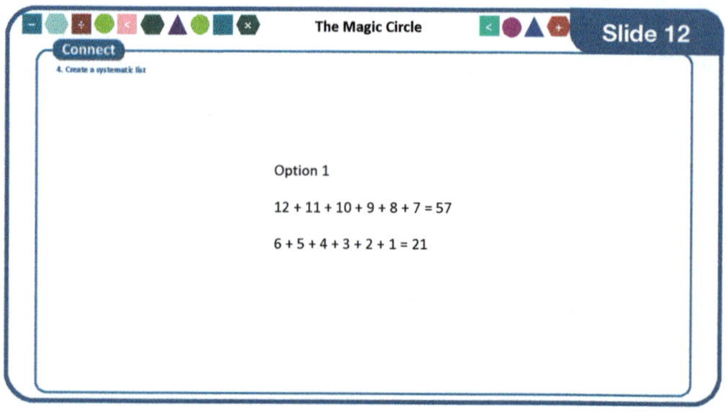

⚠ Watch for children who use this as their first try and do not consider the size of the numbers in each group and how these will not be near to equal.

- *What could we do rather than starting from scratch?*

Encourage children to improve their lists rather than starting again.

Slide 13

Display slide 13 which shows another possible list of numbers for the two rings.

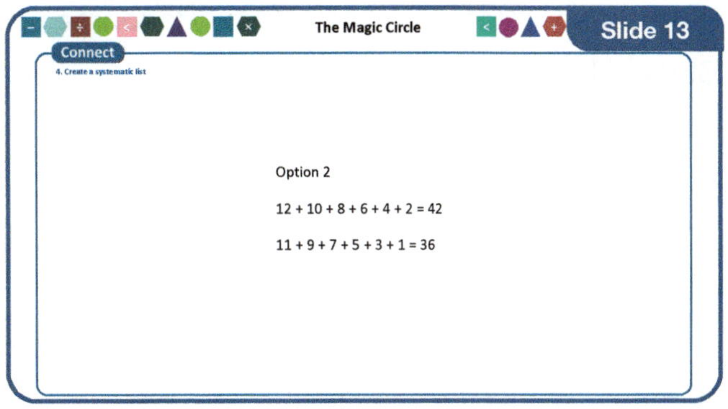

- *What has changed here?*
- *Does this satisfy the rules now?*

Encourage children to consider what they know about the difference each time.

- 42 and 36 are not equal.
- They have a difference of 6.
- I need to make one bigger by 3 and one smaller by 3.
- *What could you do next?*
- I will swap 8 and 5 because they have a difference of 3 so 42 – 3 = 39 and 36 + 3 = 39. If I swap them over in the lists, each line will be equal.

12 + 10 + (8) + 6 + 4 + 2 = ~~42~~ **39**

11 + 9 + 7 + (5) + 3 + 1 = ~~36~~ **39**

Connect

Refer to the **Collaborate** task of the **Pupil Resource**.

This task is very suitable for children to work in groups to solve the puzzle, before mixing up the groups so that children can share and compare their strategies.

Give each child in the group a number, colour or fruit name (for example). Once the groups have had sufficient time to explore the puzzle and consider solutions, ask children to regroup with other members of the same colour / number / fruit. Once regrouped, they can then discuss and share how their original group tackled the problem. Children could then return to their original groups to share strategies discussed.

Children could work together to create their explanation text as detailed in the **Communicate** task or when extending or creating their own puzzle as detailed in the **Create** task.

- *How did you decide where to start?*
- *Can you find someone who started differently? What was the difference?*

Collaborate

Refer to the **Communicate** task of the **Pupil Resource**.

This task provides an opportunity for children to write an explanation text to show how they found one or more solutions to the magic circle puzzle.

Ask children to:

- address the reader directly;
- ensure that sentences or notes are sequenced logically;
- use diagrams such as flow charts to organise the stages of the process they followed.

They can use the word bank for vocabulary support.

First	**Begin by …**	**After that …**	**When …**
At this point …	**If …, then …**	**Consequently …**	**If this works, then**
It might be advisable to …	**The final stage will involve …**	**You'll be surprised to know that …**	**Now that you know more about …**
total	**sum**	**divide**	**inner and outer rings**

Communicate

Create

Refer to the **Create** task of the **Pupil Resource**.

Slide 14

Display slide 14.

- *Were there any questions asked at the start of the task that you now want to investigate?*
- *Is there anything new that you are wondering now?*

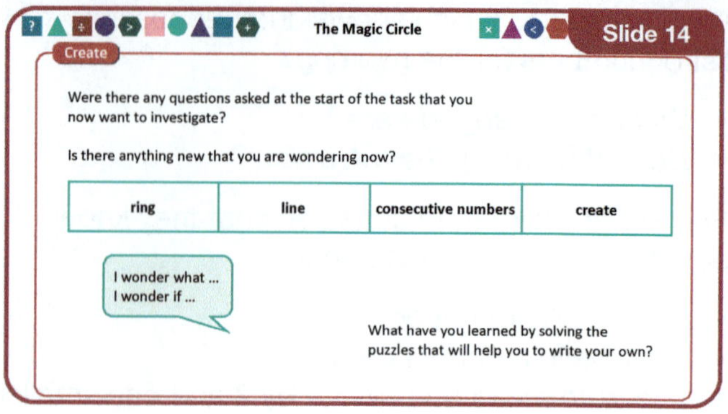

Children can use the word bank to prompt further questions of their own.

ring	**line**	**consecutive numbers**	**create**

- I wonder what would happen if we added another ring or another line.
- I wonder if it would work with the numbers 10 to 22 instead of 1 to 12.
- I wonder if this only works with consecutive numbers.
- I wonder if I could create my own similar puzzle.
- *What have you learned by solving the puzzles that will help you to write your own?*
- I will need to find the total of my chosen numbers and divide these by the number of rings and number of lines.

Slide 15

Display or print out the additional magic circles on slide 15 for children to complete. Alternatively, children could create their own magic circle.

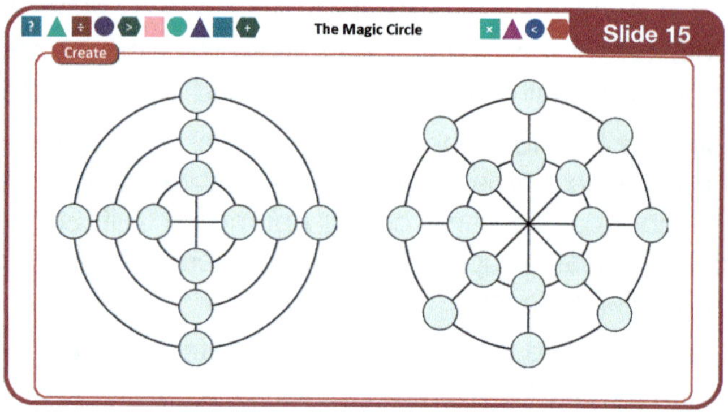

Consider

Refer to the **Consider** task of the **Pupil Resource**.

Encourage children to think about the strategies that they have used. The prompts below may help get children started.

Cognitive learning domain

- The trickiest part of solving the puzzle was …
- Someone could / could not argue with my answers because …
- The most important mathematics I learned by working on this puzzle was …

Affective domain

- Working with others on this puzzle made me feel …
- I think others felt … about working with me.
- If I could change the way we worked on the puzzle, I would …

Metacognitive domain

- I worked well on my own because …
- I interacted with others by …
- From working as part of a group, I learned …
- Next time I work as part of a group, I will …

Answers

Possible answers include:

Curious

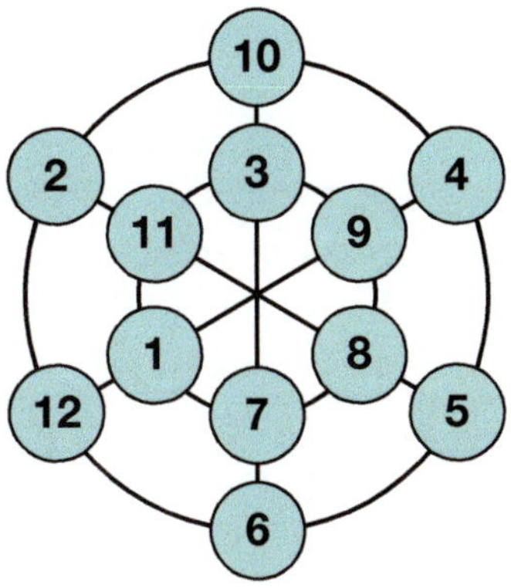

Numbers sum to 78
Rings sum to 39
Lines sum to 26

Connect: simpler puzzle

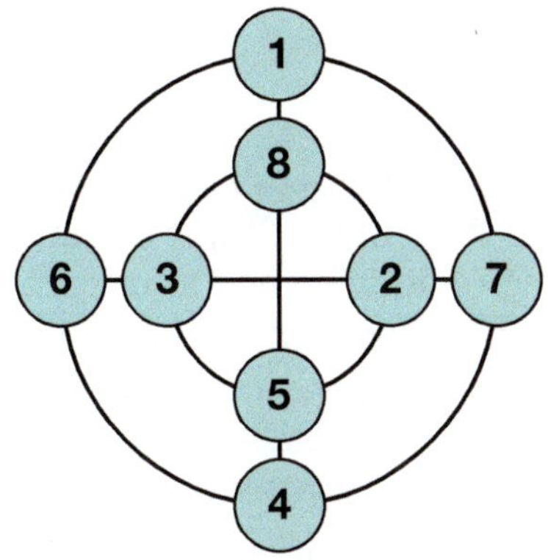

Numbers sum to 36
Rings sum to 18
Lines sum to 18

Create: puzzle 1

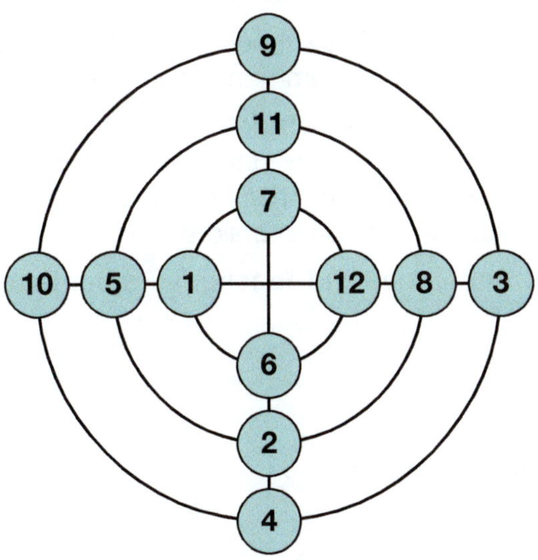

Numbers sum to 78
Rings sum to 26
Lines sum to 39

Create: puzzle 2

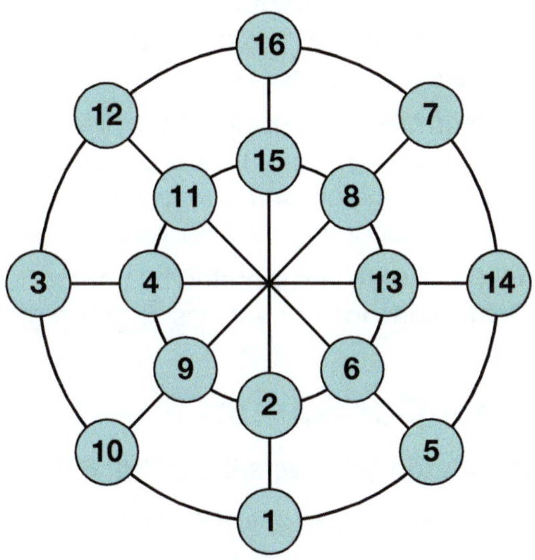

Numbers sum to 136
Rings sum to 68
Lines sum to 34

Does It All Add Up?

Year 6

Introduction

Does It All Add Up? gives children an opportunity to think more deeply about their understanding of addition and to use what they know within a more complex decimal problem. Children explore the structure of a number pyramid where each box is the sum of the two boxes directly below it.

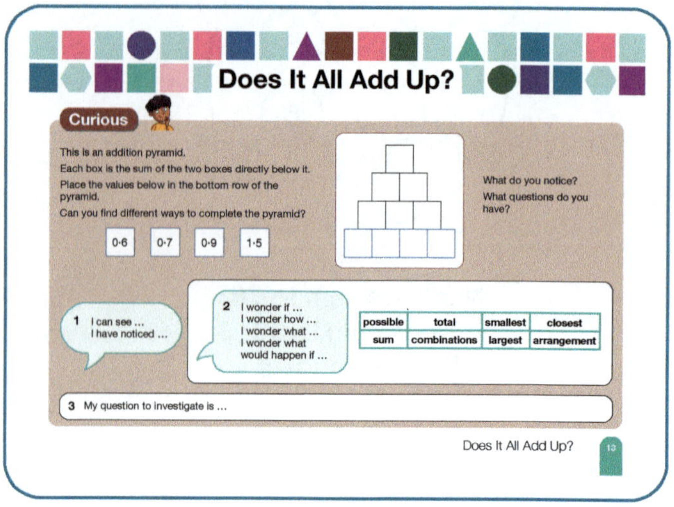

During the **Connect** stage, the focus is on children using a range of heuristics to explore all, or as many as possible, of the arrangements of the numbers, in order to work towards a generalisation. Such a generalisation will allow them to apply what they have learned to any set of numbers. Some children may attempt to present a formula or algebraic expression, explaining how and why this works. They should then also consider whether algebra is the most concise, effective and accurate method of communicating their findings.

This task promotes greater-depth learning in a range of ways, supporting children to identify and articulate underlying mathematical structures. For example, in the **Consider** stage, greater depth is promoted by asking children to identify learning behaviours and strategies that were and were not useful, and to reflect on how their learning benefited from working with other children.

This **Teacher Guide** is supported by photocopiable **Pupil Resources** and **Teacher's Presentation** PowerPoint slides for whole-class or group display.

National Curriculum

Does It All Add Up? covers these Year 6 National Curriculum objectives.

Addition, subtraction, multiplication and division

- Perform mental calculations, including with mixed operations and large numbers.

Algebra

- Use simple formulae.
- Express missing number problems algebraically.

The 6C cycle

The grid on the right shows the relative weighting given to each of the 6 Cs in **Does It All Add Up?**

Curious	★
Connect	★★★
Collaborate	★★★
Communicate	★
Create	★★
Consider	★★★

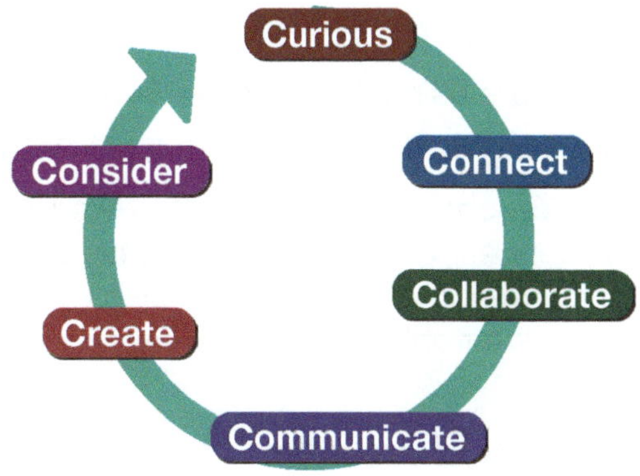

Problem-solving heuristics

The grid shows which problem-solving strategies children are likely to use when investigating the Connect tasks in **Does It All Add Up?** These strategies are further exemplified in the Connect stage below (pages 55–57).

Draw a diagram	✓
Make a table / spot a pattern	✓
Make a model / act it out	
Try then improve	
Create a systematic list	✓
Solve a simpler but related problem	✓
Work backwards	

Key mathematical ideas

Before exploring the pyramid, it is vital that children understand its structure, and that each box is the sum of the two boxes directly below it. For example:

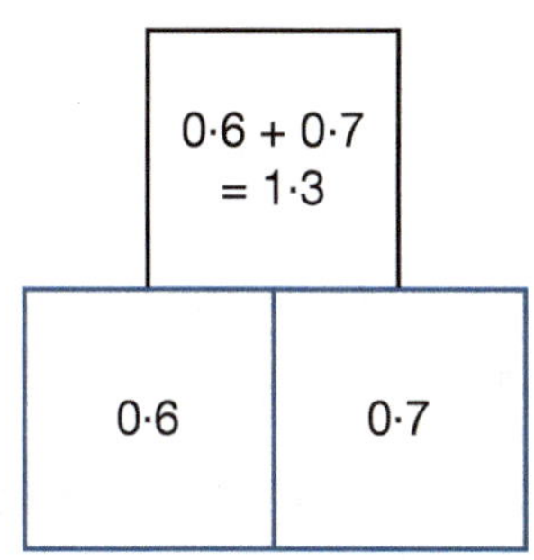

In a number pyramid, the numbers on the lower levels determine the numbers above them. Children may notice that the largest totals occur at the top of the pyramid when the number in the middle of the bottom row is the largest in that row. Conversely, others may first discover that if a smaller starting number is placed there, the number at the top of the pyramid will be smaller.

Once children have started to spot patterns and generalise, make it clear that they should justify and explain their generalisations. They should explain what is happening and why.

This model uses algebra to show how the numbers on the bottom row reoccur in the addition pyramid. In the Connect tasks a variety of models are exemplified, for example, using coloured shapes instead of numbers.

Answers are given in situ throughout the tasks, and are also summarised on page 60.

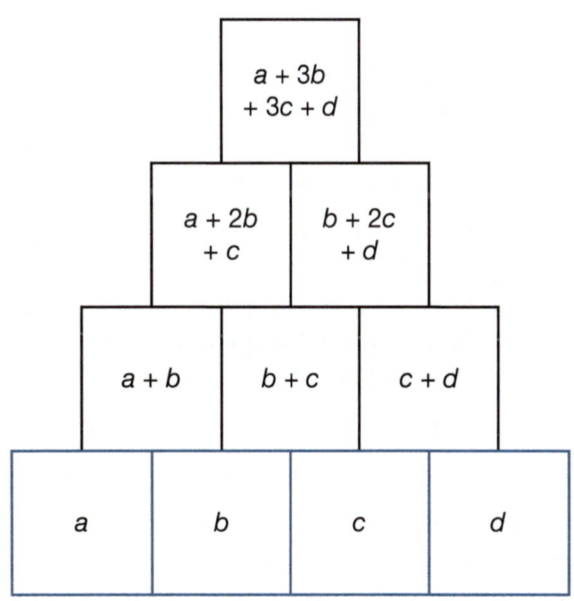

Teaching guidance

If you are using the **Express Pathway**, please turn to pages 55–59.

Teaching guidance for the **Full 6C Pathway** is given below.

The **Teacher's Presentation** PowerPoint file has slides specifically mentioned in this teaching guidance. Some of the slides include notes (within the PowerPoint file) providing 'stumble support' suggestions, that may help you to support children in their problem-solving thinking and enable access for children who are less confident.

Awe and wonder

Reveal slides 1–2 one at a time, allowing time for discussion and observations as prompted below.

Slide 1

Display slide 1.

Allow children to share what they are noticing or wondering.

- I have noticed that the pyramid has four rows.
- I have noticed that there are a total of 10 boxes.
- I can see that each row has one more box than the row above.

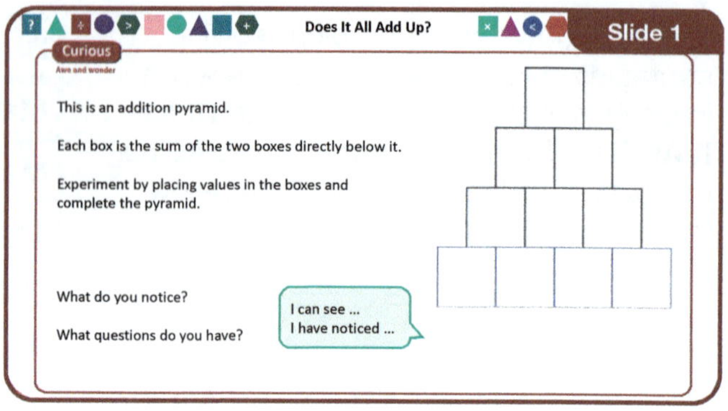

Slide 2

Display slide 2.

Allow children to share anything new they are now noticing or wondering.

- I can see that all four numbers have one decimal place.

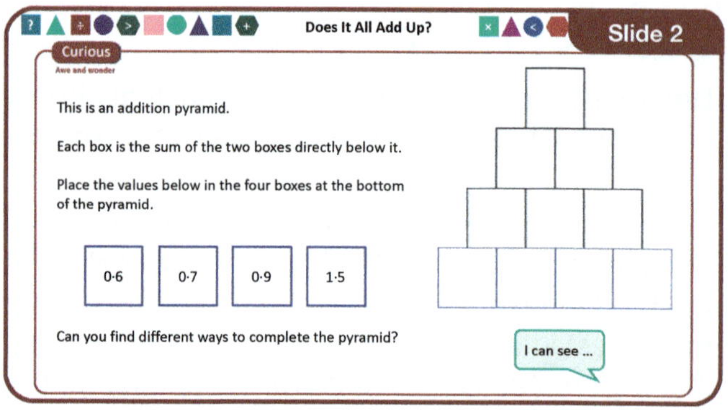

Develop and deepen

Encourage children to use the words in the word bank to develop and deepen their thoughts and questions.

possible	**total**	**smallest**	**closest**
sum	**combinations**	**largest**	**arrangement**

Possible lines of enquiry linked to the word bank:

- I wonder how many combinations there are for arranging the bottom row of numbers.
- I wonder how the arrangement of the bottom row impacts each row of the pyramid.
- I wonder what the largest / smallest possible total for the top of the pyramid is.
- I wonder if there is a quick way of calculating the top of the pyramid without calculating every box in the pyramid.

Possible lines of enquiry that children may raise now, but should be held back until the **Create** task:

- I wonder what would happen if I increased the number of rows to create larger pyramids.
- I wonder what would happen if I multiplied the numbers instead of adding them.

Consider whether / how to collate and display all the suggested questions, and whether / how to steer children towards selecting an appropriately challenging question.

Pupil Resources for Full and Express Pathways

Distribute the **Pupil Resources** for **Does It All Add Up?** to the children to help support them in their exploration of the different tasks. Use either of the **Curious Pupil Resources** depending on whether you are following the **Full Pathway** or the **Express Pathway**. The **Pupil Resources** for the remaining 5Cs are common to both pathways.

To provide additional flexibility in the classroom, each of the **Pupil Resources** is also provided in the **Teacher's Presentation** PowerPoint file (see Slides 3–6), so you can display these at any time if you wish.

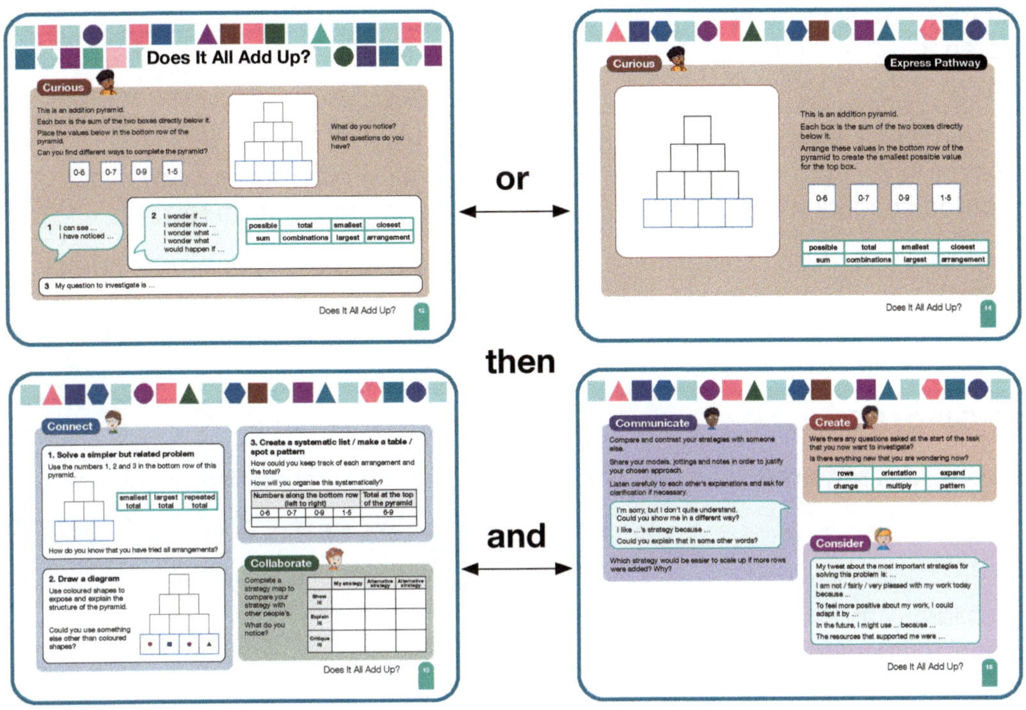

Express Pathway

If you are following the **Express Pathway**, spend a few minutes discussing the **Curious Pupil Resource** with the children. (This is slide 4, so you could display it if you wish.) Use some of the prompts on the previous page to help.

Three tasks involving different strategies are suggested for children to make connections.

1. Solve a simpler but related problem

Slide 7

Display slide 7, which shows a three-row pyramid and a simpler number range for children to investigate.

- *Using the numbers 1, 2 and 3 in the bottom row of the pyramid, what can you find out?*

Encourage children to use the sentence frames to talk about their findings.

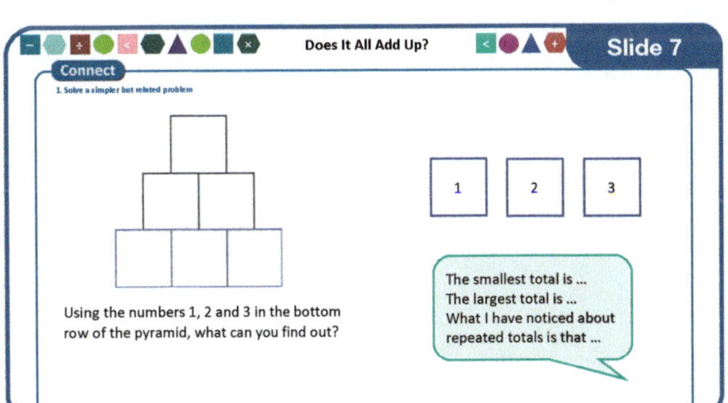

- The smallest total is 7.
- There are three different totals, 7, 8 and 9 and each are repeated twice.

Slide 8

Display slide 8. Discuss the different pyramids that are possible when starting with numbers 1, 2 and 3 on the bottom row.

- *Have all combinations been shown here? How do you know?*
- *What do you notice about the way that these arrangements have been organised?*

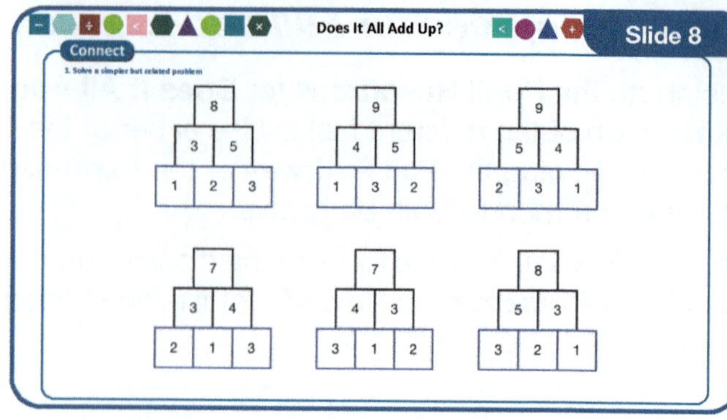

Working with these smaller examples may support children to make the following observations:

- Some combinations are repeated as they are symmetrical.
- The smallest total occurs when 1 is in the middle number on the bottom row.
- The largest total occurs when 3 is in the middle number on the bottom row.

Children could then use these findings to see how they relate to the original, more complex four-row decimal pyramid.

2. Draw a diagram

Additional resources required: counters or interlocking cubes in four different colours (optional)

Slide 9

Display slide 9.

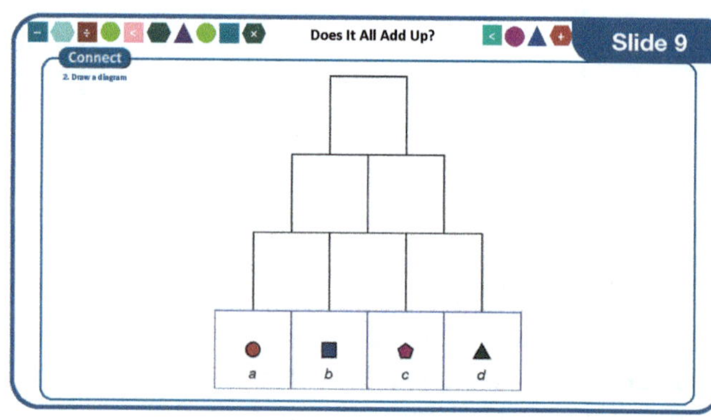

Explain that children can draw a diagram or build a model to uncover and explain the structure of the pyramid. Removing the numbers from the pyramid will support children to focus on the structure. Children may choose to replace the numbers with colours, shapes, symbols or letters.

This could be supported by using coloured counters or cubes to build a physical model.

Slide 10

Display slide 10, which provides an example of how colours and shapes could be used to show how the numbers in the bottom reoccur throughout the addition pyramid. You may need to explain what 2*b*, for example, means and where it comes from, explaining that this is the algebraic way of writing $2 \times b$ ($2 \times$ blue squares, in this case).

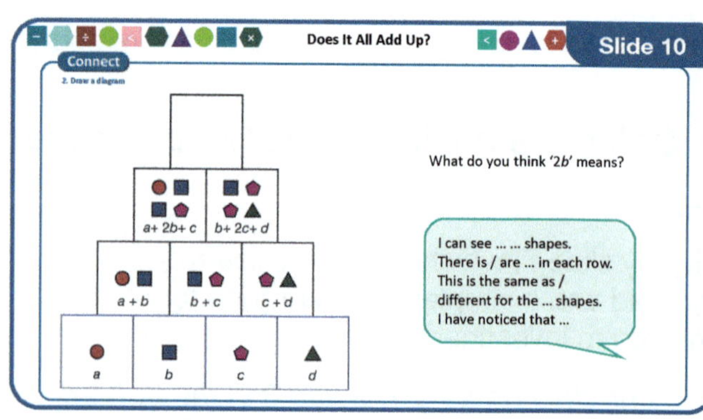

Encourage children to look at how many of each coloured shape there are in each row of the diagram and to use the sentence frames on the slide to describe what they see.

- I can see 4 red circles in total. There is 1 in each row. This is the same for the green triangles.
- I can see 9 blue squares in total. There is 1 in the bottom row, then 2 then 3, then 3. I have noticed that if the shapes are in boxes next to each other, then these are combined in the box above.

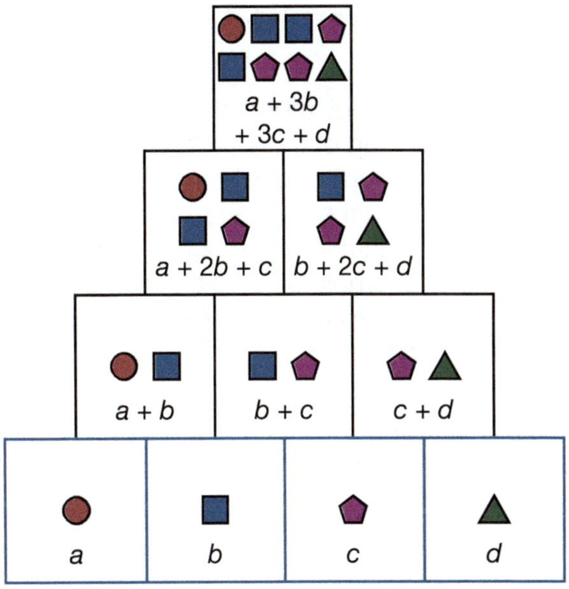

If children notice patterns but find it difficult to explain them, it may be helpful to support them in using the algebraic representation.

- *Can you explain any generalisations that you have reached?*
- *Can you use algebra to explain your thinking?*

3. Create a systematic list / make a table / spot a pattern

Slide 11

Display slide 11, which shows a partially completed table.

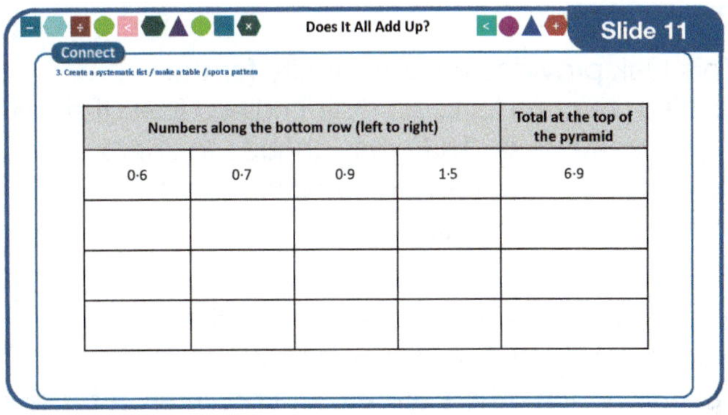

Ask children to make a systematic list of the arrangements by copying and completing the table. There are 24 distinct combinations which may take a long time to work through. After children have completed several rows, they may start to spot patterns. Whilst some children may choose to complete the full table, others may choose to use what they have learned from the systematic list so far to start making generalisations.

This table shows all the combinations for when 0·6 is the first number on the left.

Numbers along the bottom row (left to right)				Total at the top of the pyramid
0·6	0·7	0·9	1·5	6·9
0·6	0·7	1·5	0·9	8·1
0·6	0·9	0·7	1·5	6·9
0·6	0·9	1·5	0·7	8·5
0·6	1·5	0·9	0·7	8·5
0·6	1·5	0·7	0·9	8·1

When looking closely, it is possible to see that if the outside numbers are kept the same, the total at the top of the pyramid remains the same, regardless of the position of the middle two numbers.

Others may notice that the higher totals at the top of the pyramid occur when the largest values occur in the two middle boxes.

Collaborate

Refer to the **Collaborate** task of the **Pupil Resource.**

Slide 12

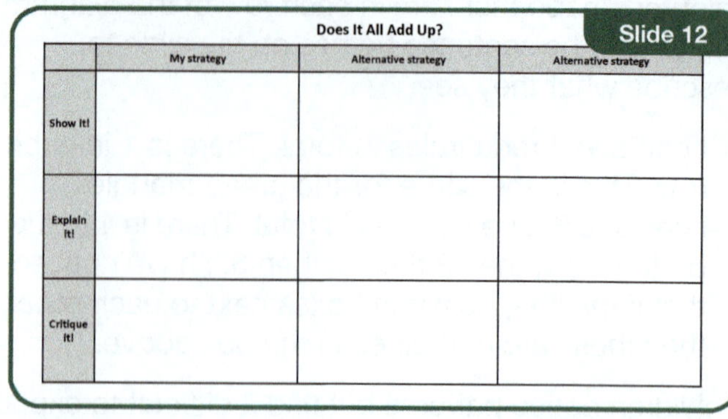

Print slide 12 and distribute the strategy map to each pair or group of three children. Organise children into groups according to the way they represented the numbers in the second **Connect** task, that is, using colours, symbols or letters. The template could be adapted to include more or fewer strategies, or children may design their own.

Pairs should collaborate to prepare a presentation about their chosen strategy, before taking it in turns to communicate the strategies. Ask each child to make notes about the variety of methods for their own records on their strategy map.

Children will need to collaborate closely to decide how best to articulate and explain their chosen strategy so that their audience will understand and be able to summarise it on their own strategy map.

Communicate

Refer to the **Communicate** task of the **Pupil Resource.**

Slide 13

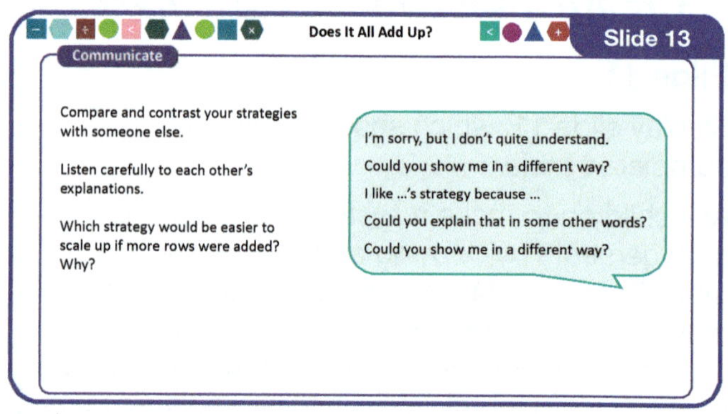

Display slide 13.

This task provides an opportunity for the children to communicate their findings from the **Collaborate** task. They may choose to share their models, jottings and notes in order to justify their chosen approach.

These presentations allow children to compare and contrast the different strategies that have been used. Encourage children to listen carefully when others are presenting, interpret their presentations and seek clarification where necessary.

- I'm sorry, but I don't quite understand. Could you show me in a different way?
- I like …'s strategy because …
- *Which strategy would be easier to scale up if more rows were added? Why?*

Refer to the **Create** task of the **Pupil Resource**.

Slide 14

Display slide 14.

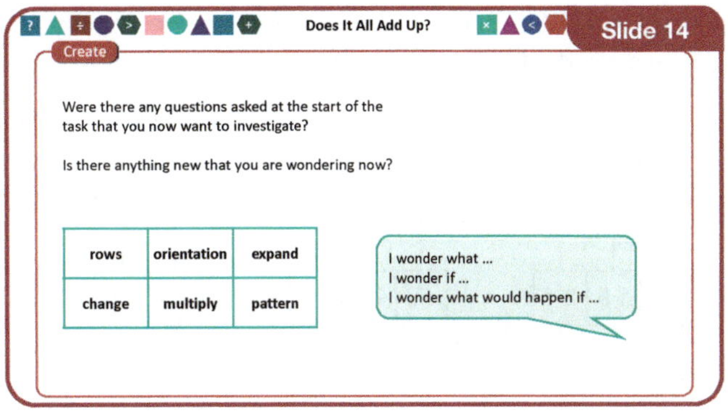

- *Were there any questions asked at the start of the task that you now want to investigate?*
- *Is there anything new that you are wondering now?*

Children can use the word bank to prompt further questions of their own.

rows	**orientation**
expand	**change**
multiply	**pattern**

- I wonder what would happen if I increased the number of rows to create larger pyramids.
- I wonder what would happen if I multiplied the numbers instead of adding them.
- I wonder if I could start with fractions or percentages instead of decimal numbers.
- I wonder if I can design a different shape for an addition puzzle.

Refer to the **Consider** task of the **Pupil Resource**.

Slide 15

Display slide 15.

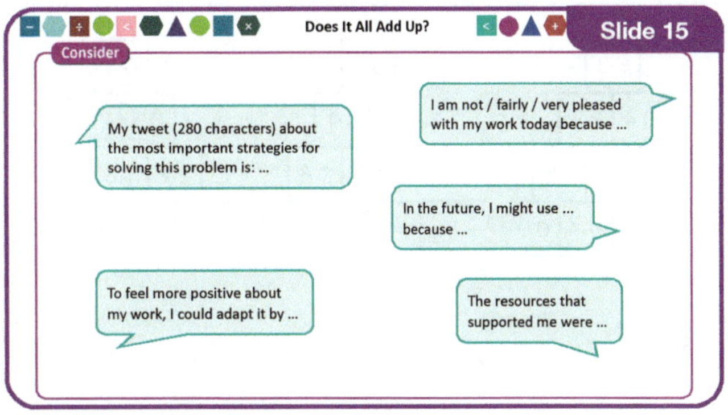

Encourage children to think about the strategies that they have used. The prompts below may help get children started.

Cognitive learning domain

- My tweet about the most important strategies for solving this problem is: …
- My tweet is / is not similar to others in the class. I think that this is because …
- I think that some children might still struggle to understand the answer to this problem because …
- To help them understand, I could …

Affective domain

- I am not / fairly / very pleased with my work today because …
- The way I worked was …
- In the future, I might use ... because …
- To feel more positive about my work, I could adapt it by …

Metacognitive domain

- The resources that supported me were …
- These supported me by …
- … was less helpful because … Because of this, I …

Answers

There are 24 different ways to order the four numbers in the bottom row, and therefore 24 different pyramids that can be created.

There are six possible totals for the top of the pyramid: 6·3, 6·7, 6·9, 7·9, 8·1 and 8·5.

6·9
2·9 | 4
1·3 | 1·6 | 2·4
0·6 | 0·7 | 0·9 | 1·5

8·1
3·5 | 4·6
1·3 | 2·2 | 2·4
0·6 | 0·7 | 1·5 | 0·9

6·9
3·1 | 3·8
1·5 | 1·6 | 2·2
0·6 | 0·9 | 0·7 | 1·5

8·5
3·9 | 4·6
1·5 | 2·4 | 2·2
0·6 | 0·9 | 1·5 | 0·7

8·1
4·3 | 3·8
2·1 | 2·2 | 1·6
0·6 | 1·5 | 0·7 | 0·9

8·5
4·5 | 4
2·1 | 2·4 | 1·6
0·6 | 1·5 | 0·9 | 0·7

6·7
2·8 | 3·9
1·3 | 1·5 | 2·4
0·7 | 0·6 | 0·9 | 1·5

7·9
3·4 | 4·5
1·3 | 2·1 | 2·4
0·7 | 0·6 | 1·5 | 0·9

6·7
3·1 | 3·6
1·6 | 1·5 | 2·1
0·7 | 0·9 | 0·6 | 1·5

8·5
4 | 4·5
1·6 | 2·4 | 2·1
0·7 | 0·9 | 1·5 | 0·6

7·9
4·3 | 3·6
2·2 | 2·1 | 1·5
0·7 | 1·5 | 0·6 | 0·9

8·5
4·6 | 3·9
2·2 | 2·4 | 1·5
0·7 | 1·5 | 0·9 | 0·6

6·3
2·8 | 3·5
1·5 | 1·3 | 2·2
0·9 | 0·6 | 0·7 | 1·5

7·9
3·6 | 4·3
1·5 | 2·1 | 2·2
0·9 | 0·6 | 1·5 | 0·7

6·3
2·9 | 3·4
1·6 | 1·3 | 2·1
0·9 | 0·7 | 0·6 | 1·5

8·1
3·8 | 4·3
1·6 | 2·2 | 2·1
0·9 | 0·7 | 1·5 | 0·6

7·9
4·5 | 3·4
2·4 | 2·1 | 1·3
0·9 | 1·5 | 0·6 | 0·7

8·1
4·6 | 3·5
2·4 | 2·2 | 1·3
0·9 | 1·5 | 0·7 | 0·6

6·3
3·4 | 2·9
2·1 | 1·3 | 1·6
1·5 | 0·6 | 0·7 | 0·9

6·7
3·6 | 3·1
2·1 | 1·5 | 1·6
1·5 | 0·6 | 0·9 | 0·7

6·3
3·5 | 2·8
2·2 | 1·3 | 1·5
1·5 | 0·7 | 0·6 | 0·9

6·9
3·8 | 3·1
2·2 | 1·6 | 1·5
1·5 | 0·7 | 0·9 | 0·6

6·7
3·9 | 2·8
2·4 | 1·5 | 1·3
1·5 | 0·9 | 0·6 | 0·7

6·9
4 | 2·9
2·4 | 1·6 | 1·3
1·5 | 0·9 | 0·7 | 0·6

Dice It Up

Year 5

Introduction

Dice It Up asks children to investigate the number of dots on the outside of a cuboid made from a stack of dice. The dice in the cuboid are all set in the same orientation. Only three faces of the cuboid are visible in the picture. Before you start, give children some time to explore an individual dice, and in particular the fact that the dots on opposite sides of a dice total 7. This information is key as children will need to visualise the faces they cannot see, based on those they can.

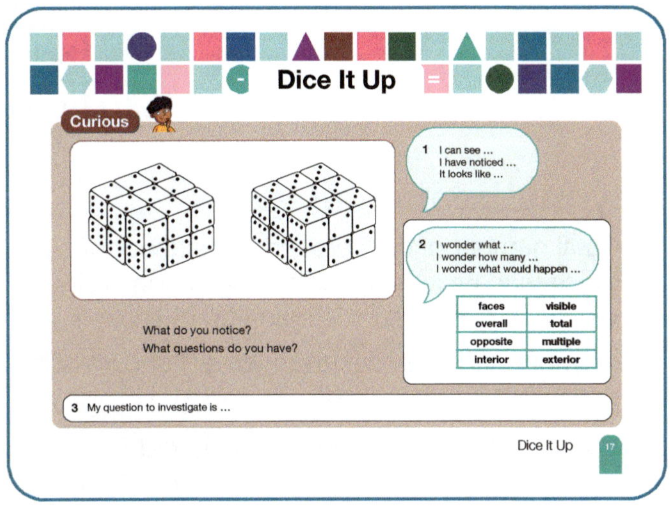

There are a number of problem-solving heuristics to choose from involving systematic thinking and organisation as well as application of calculation strategies and recall of number facts. Once children have answered their own initial questions about the dice cuboids, they are supported to generalise about their findings.

This task promotes greater-depth learning by supporting children to identify and articulate underlying mathematical structures. For example, in the **Curious** stage, greater depth is promoted by asking children to make rich mathematical contributions to discussions and suggest a variety of appropriate mathematical questions.

This **Teacher Guide** is supported by photocopiable **Pupil Resources** and **Teacher's Presentation** PowerPoint slides for whole-class or group display.

National Curriculum

Dice It Up covers these Year 5 National Curriculum objectives.

Multiplication and division

- Solve problems involving addition, subtraction, multiplication and division and a combination of these, including understanding the meaning of the equals sign.

The 6C cycle

The grid on the right shows the relative weighting given to each of the 6 Cs in **Dice It Up**.

Curious	★★★
Connect	★★
Collaborate	★
Communicate	★★
Create	★★
Consider	★★

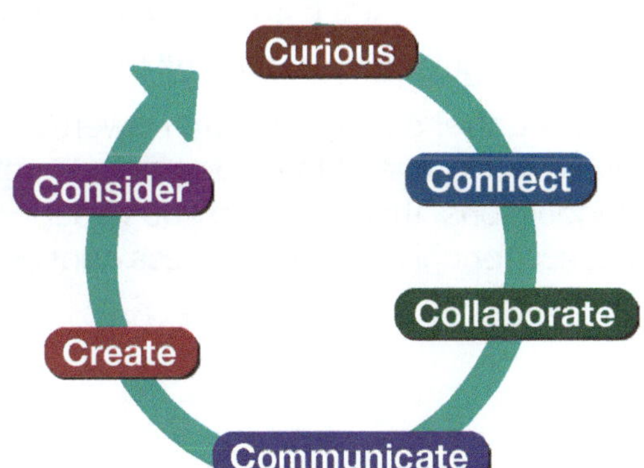

Problem-solving heuristics

The grid shows which problem-solving strategies children are likely to use when investigating the **Connect** tasks in **Dice It Up**. These strategies are further exemplified in the **Connect** stage below (pages 64–67).

Draw a diagram	✓
Make a table / spot a pattern	✓
Make a model / act it out	✓
Try then improve	
Create a systematic list	
Solve a simpler but related problem	✓
Work backwards	

Key mathematical ideas

The total number of dots around the exterior of a $3 \times 3 \times 2$ cuboid built from dice that are all in the same orientation is always 147. In a $3 \times 3 \times 2$ cuboid, each cuboid face is made up of either 6 or 9 dice faces.

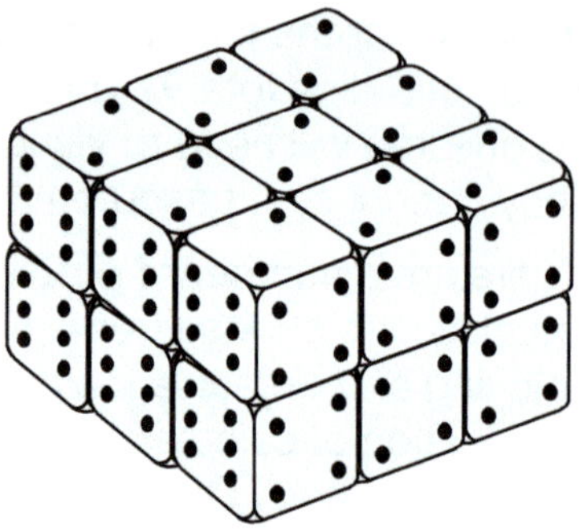

The number of dots in the arrangement shown here can be calculated as:

number of dots on the three visible cuboid faces:

$$(6 \times 6) + (6 \times 4) + (9 \times 2) = 78$$

+ number of dots on the three hidden cuboid faces:

$$(6 \times 1) + (6 \times 3) + (9 \times 5) = 69$$

$$= 147$$

Because we know that opposite faces on a dice total 7, we can shorten this calculation to:

number of dots on all the exterior cuboid faces: $(6 \times 7) + (6 \times 7) + (9 \times 7) = 147$

The brackets are not needed due to the law of BIDMAS but are used here to make the groupings clearer and easier to relate to the model.

Because the opposite sides of a dice always add up to 7, alternative arrangements also yield the same number of dots, regardless of which dice faces are visible (as long as all the dice in the arrangement are all in the same orientation).

Answers are given in situ throughout the tasks because of the cumulative nature of the tasks.

Teaching guidance

If you are using the **Express Pathway**, please turn to pages 64–69.

Teaching guidance for the **Full 6C Pathway** is given below.

The **Teacher's Presentation** PowerPoint file has slides specifically mentioned in this teaching guidance. Some of the slides include notes (within the PowerPoint file) providing 'stumble support' suggestions, that may help you to support children in their problem-solving thinking and enable access for children who are less confident.

Awe and wonder

Slide 1

Display slide 1.

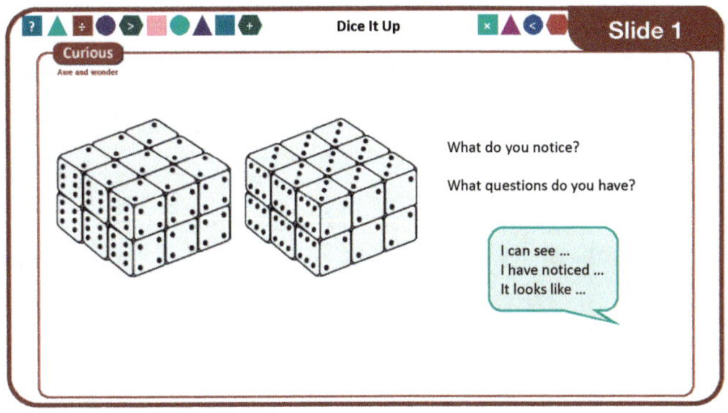

Allow children to share what they are noticing or wondering.

- I can see lots of dots.
- I have noticed that the dice have been made into cuboids.
- I can see that there are groups of dice showing the same number.
- It looks like the dice are all facing the same way.
- It looks like there are more dots in the first picture than the second.

Develop and deepen

Encourage children to use the words in the word bank to develop and deepen their thoughts and questions.

faces	**opposite**	**visible**	**total**
overall	**interior**	**exterior**	**multiple**

Possible lines of enquiry:

- I wonder what the total of all the visible dots is.

Possible lines of enquiry linked to the word bank:

- I wonder what the overall number of dots around the outside would be.
- I wonder what the dots of opposite sides of a dice add up to.
- I wonder how many dots are hidden.

Possible lines of enquiry that children may raise now, but should be held back until the **Create** task:

- I wonder how many dots there would be if the cuboid had more layers.
- I wonder how many dots there would be on the outside of a 3 by 3 by 3 dice cube.

Consider whether / how to collate and display all the suggested questions, and whether / how to steer children towards selecting an appropriately challenging question.

Pupil Resources for Full and Express Pathways

Distribute the **Pupil Resources** for **Dice It Up** to the children to help support them in their exploration of the different tasks. Use either of the **Curious Pupil Resources** depending on whether you are following the **Full Pathway** or the **Express Pathway**. The **Pupil Resources** for the remaining 5Cs are common to both pathways.

To provide additional flexibility in the classroom, each of the **Pupil Resources** is also provided in the **Teacher's Presentation** PowerPoint file (see slides 2–5), so you can display these at any time if you wish.

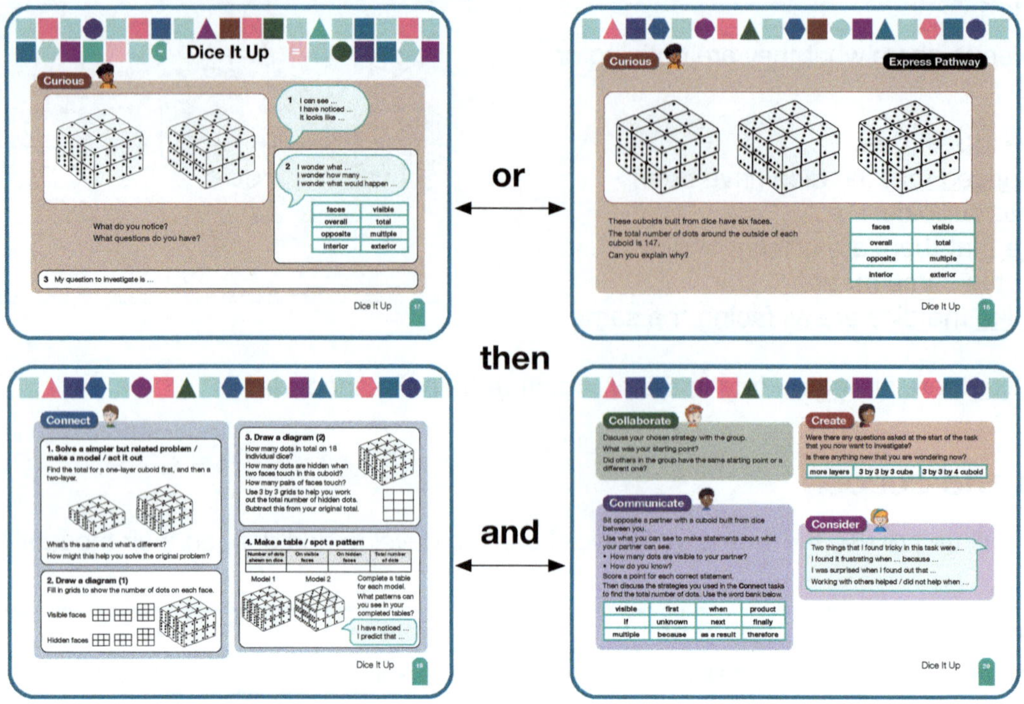

Express Pathway

If you are following the **Express Pathway**, spend a few minutes discussing the **Curious Pupil Resource** with the children. (This is slide 3, so you could display it if you wish.) Use some of the prompts on the previous page to help.

Four tasks involving different strategies are suggested for children to make connections.

Additional resources required: dice (1–6)

For all the **Connect** tasks, make dice available to children so that they can explore the arrangements of dots in opposite faces and the resulting number patterns.

Ask children to look closely at one of the dice, paying attention to the opposite sides. Agree that opposite sides of a dice have 7 dots altogether.

1. Solve a simpler but related problem / make a model / act it out

Slide 6

Display slide 6. Ask children to first investigate a cuboid with one layer of dice. Children could build the layer of dice shown in the image.

- *What can you see?*
- I can see 3 groups of 6 dots on this face, 3 groups of 4 dots on this face and 9 groups of 2 dots on the top face. So that's 18 + 12 + 18 = 48 dots.

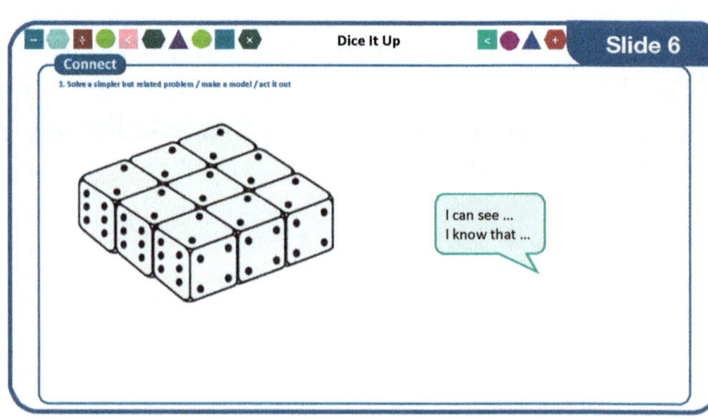

- *How could you work out the number of dots that you can't see?*
- Opposite sides total 7. So, on the hidden faces, there must be 3 groups of 1 dot, 3 groups of 3 dots and 9 groups of 5 dots. That's $3 + 9 + 45 = 57$ dots.

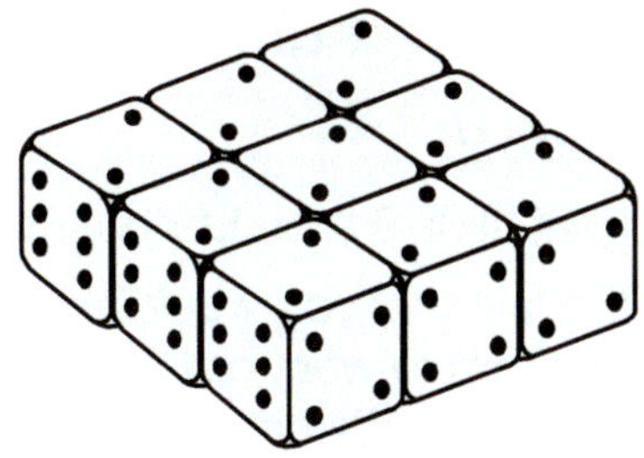

Agree that $48 + 57 = 105$, so there are 105 dots altogether.

Slide 7

Display slide 7.

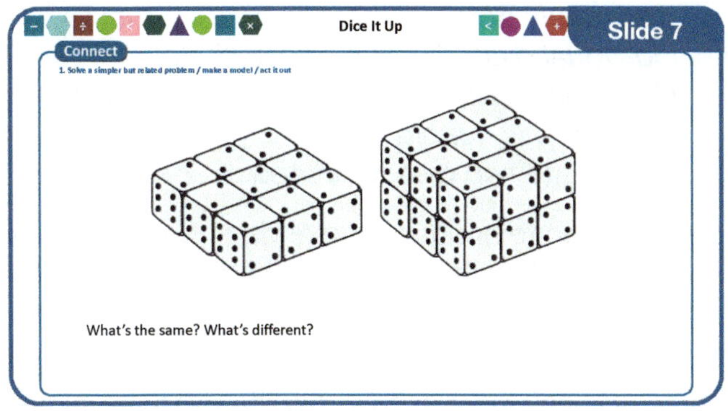

Children may rush to say that there must be 210 dots (double 105) on the two-layer cuboid. If so, remind them that in between the two layers, there are now some completely hidden dots that are not on the outside of the cuboid.

- *What do you know from the one-layer cuboid that can help you here?*
- I know that there are 9 groups of 5 dots on the opposite side of the top layer. 9×5 is 45, so there are 45 dots hidden underneath the top layer.
- 9 lots of 2 dots are hidden on top of the bottom layer. 9×2 is 18. $45 + 18$ is 63.
- $210 - 63 = 147$. So there are 147 dots altogether on the exterior faces.
- *Is there another way to calculate the number of dots hidden in the middle layer of the cuboid?*
- There will always be 63 dots hidden in the middle layer because there are 9 sets of dice touching and the sum of the touching sides will always be 7.

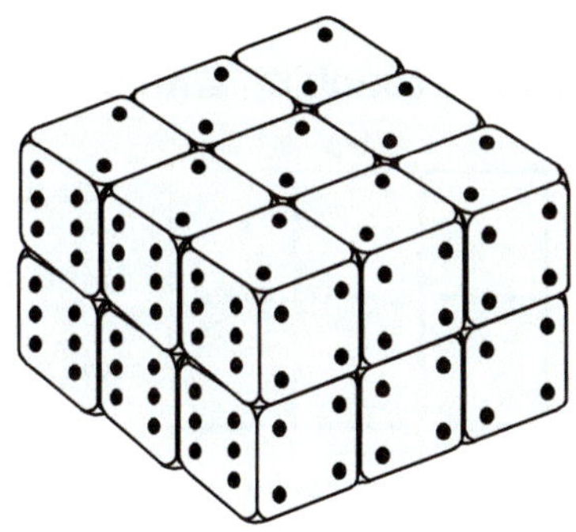

2. Draw a diagram (1)

Additional resources required: cm-squared paper

Slide 8

Display slide 8.

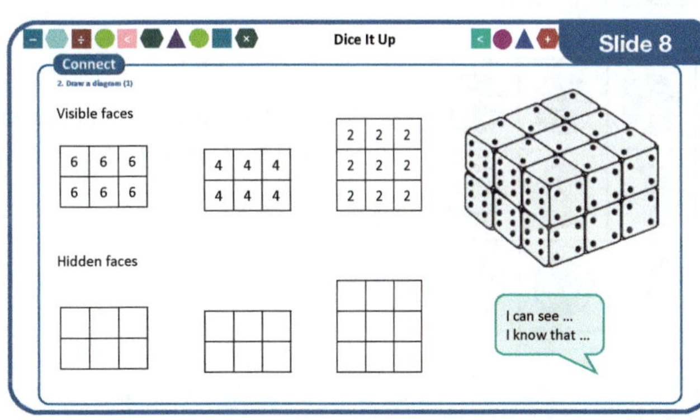

Ask children to draw a series of grids on squared paper to represent the sides of the dice cuboid that cannot be seen. Complete the grids on the slide together, showing 6×1, 6×3 and 9×5.

1	1	1
1	1	1

3	3	3
3	3	3

5	5	5
5	5	5
5	5	5

Completed grids for the hidden faces

3. Draw a diagram (2)

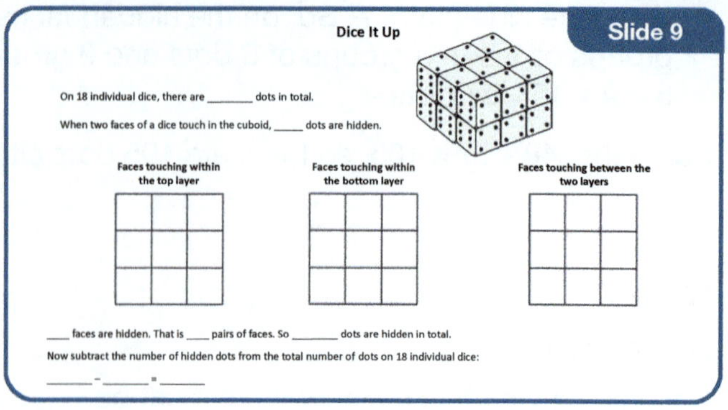

Slide 9

Print slide 9 and distribute to the children.

Explain that you would like them to use a diagram to work out the total number of dots in a slightly different way.

- *How many dots are there in total on the 18 individual dice?* ($18 \times 7 \times 3 = 378$ dots)
- *How many dots are hidden when two faces touch in this cuboid?*
- Every time two dice faces touch, 7 dots are hidden.

Ask children to annotate the grids to help them count the number of touching faces, both within each layer (the vertical faces of the dice that are touching) and between the two layers (the horizontal faces of the dice that are touching). They can then work out the number of hidden dots.

They may wish to mark the internal grid lines in different colours to show the touching faces within a layer, and draw circles to show the touching faces between the layers.

Faces touching within the top layer

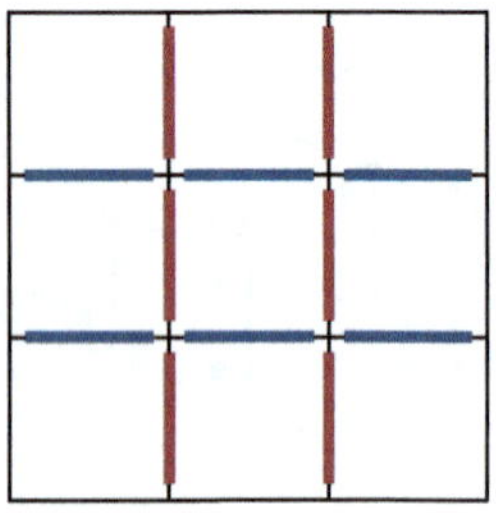

Each coloured line indicates a pair of touching vertical faces.

Red lines: 6-dot and 1-dot

Blue lines: 4-dot and 3-dot

12 pairs of touching faces

$12 \times 7 = 84$ hidden dots

Faces touching within the bottom layer

The top and bottom layers are identical, so there are 84 hidden dots for the bottom layer too.

Faces touching between the two layers

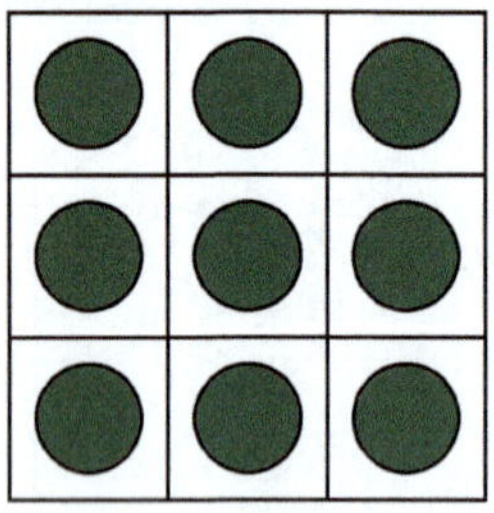

Each green circle indicates a pair of touching horizontal 2-dot and 5-dot faces.

9 pairs of touching faces

$9 \times 7 = 63$ hidden dots between the layers

Total numbers of hidden dots

$(2 \times 84) + 63 = 231$ dots hidden inside the cuboid

378 (total number of dots on all 18 dice) – 231 hidden dots

$= 147$ dots on the exterior of the cuboid

Some children may wish to annotate their grids with numbers to show the numbers of dots.

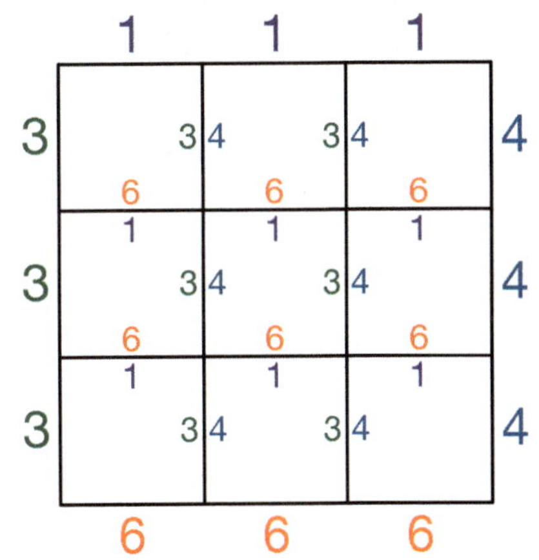

Alternative example of an annotated grid showing the touching vertical faces within a layer

4. *Make a table / spot a pattern*

Slide 10

Print slide 10 and distribute to children.

Ask them to complete a table for each of the models to work out the total number of dots on the outside of the cuboid.

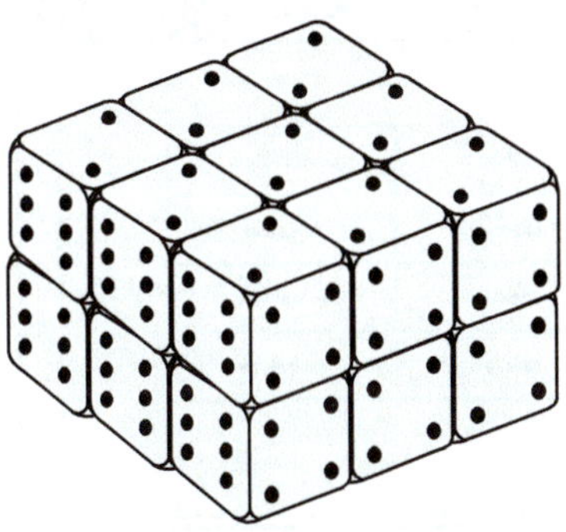

Number of dots shown on dice	On visible faces	On hidden faces	Total number of dots
1		6	6
2	9		18
3		6	18
4	6		24
5		9	45
6	6		36
		Grand total:	147

Completed table for model 1

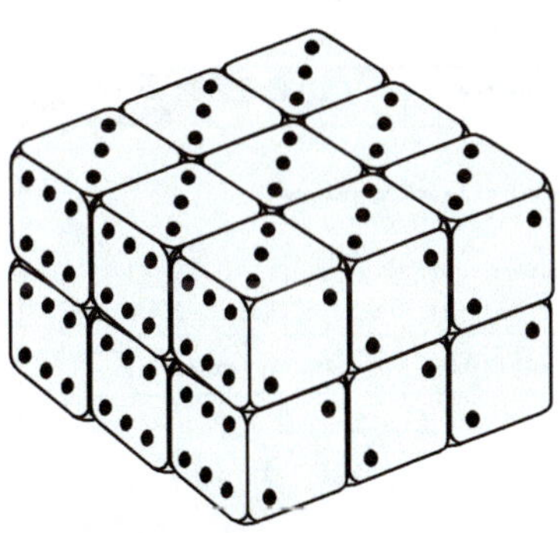

Number of dots shown on dice	On visible faces	On hidden faces	Total number of dots
1		6	6
2	6		12
3	9		27
4		9	36
5		6	30
6	6		36
		Grand total:	147

Completed table for model 2

- *What patterns can you see?*
- *What do you notice about the totals for the two different models?*
- The total of dots on the outside of a 3 by 3 by 2 dice cuboid is always 147, regardless of the dots that are shown on the top of the cuboid.

Collaborate

Refer to the **Collaborate** task of the **Pupil Resource**.

This task provides children with an opportunity to work independently to solve the problem using their chosen strategy. Ask children to reflect on their starting point, and then discuss the problem as a group.

- *Did you all start from the same starting point or did others in the group have a different one?*
- *Has this discussion made you realise anything new?*

Strategies will then be shared and reflected upon through the **Communicate** task.

Communicate

Refer to the **Communicate** task of the **Pupil Resource**.

This task provides an opportunity for the children to use what they have discovered about the opposite sides of a dice in the **Connect** tasks in a pairs game.

Ask children to sit in pairs opposite each other with a cuboid made of dice between them. Each child should only be able to see some of the faces – the faces that are hidden to Player A are the ones visible to Player B and vice versa. The bottom face of the cuboid will be hidden from both players.

Each player takes it in turns to use what they can see to make statements about what their partner can see. For each correct statement, they score a point. How many points can they score?

Following the game ask pairs to compare their strategies from the **Connect** stage.

Slide 11

Display slide 11 which shows the word bank that they can use for vocabulary support.

visible	**product**	**unknown**	**multiple**
first	**next**	**if**	**because**
when	**finally**	**therefore**	**as a result**

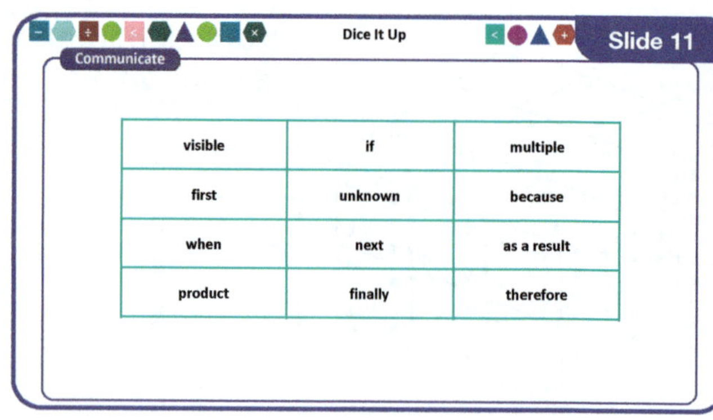

Create

Refer to the **Create** task of the **Pupil Resource**.

Slide 12

Display slide 12.

- *Were there any questions asked at the start of the task that you now want to investigate?*
- *Is there anything new that you are wondering now?*

Children can use the word bank to prompt further questions of their own.

more layers	**3 by 3 by 3 cube**	**3 by 3 by 4 cuboid**

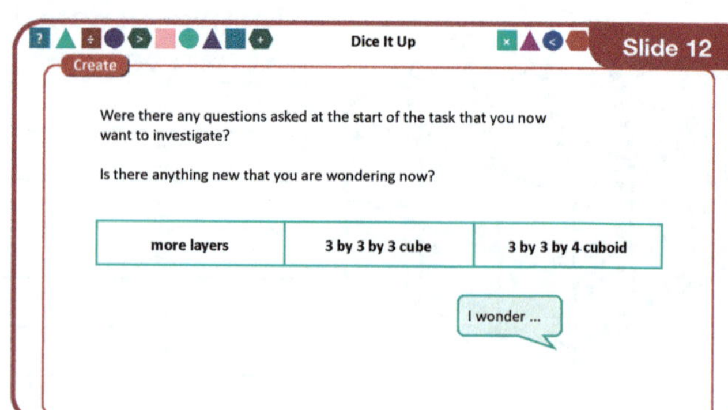

- I wonder how many dots there would be if the cuboid had more layers.
- I wonder how many dots there would be on the outside of a 3 by 3 by 3 dice cube.
- I wonder how many dots there would be on the outside of a 3 by 3 by 4 dice cuboid.

Refer to the **Consider** task of the **Pupil Resource**.

Slide 13

Display slide 13.

Encourage children to think about the strategies that they have used. The prompts below may help get children started.

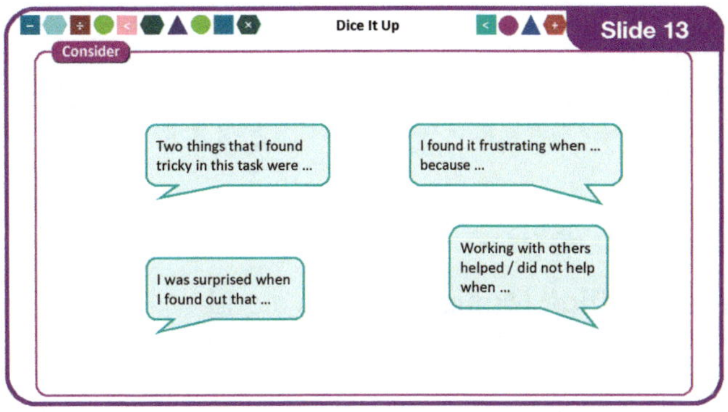

Cognitive learning domain

- Two things that I found tricky in this task were …
- The best strategy was …

Affective domain

- I found it frustrating when … because …
- I was surprised when I found out that …

Metacognitive domain

- Working with others helped / did not help when …
- Next time, I will …

Missing Digit Multiplication

Year 5
Year 6

Introduction

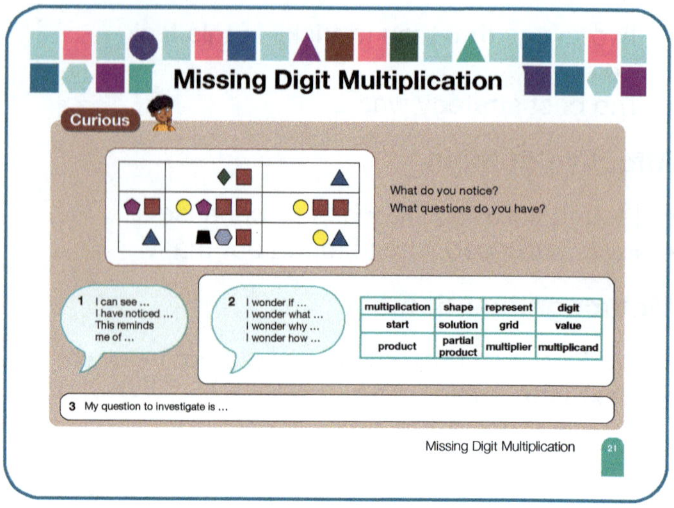

Missing Digit Multiplication is a grid multiplication puzzle where coloured shapes represent digits. If your calculation policy does not include the grid method, children could compare it to the formal written method, which is used in the **Express Pathway**.

Children should work strategically, rather than simply using trial and improvement. Encourage them to choose a suitable starting place, and then use logic to work through the puzzle, disregarding digits that do not work. Systematically creating a list or table will help children keep track of their progress.

This task promotes greater-depth learning in a range of ways, supporting children to identify and articulate underlying mathematical structures. For example, in the **Collaborate** stage, greater depth is promoted by asking children to offer suggestions, listen carefully to others and build upon their own ideas. They will also learn to encourage and support others, and be able to accept it if their ideas are rejected by the group.

This **Teacher Guide** is supported by photocopiable **Pupil Resources** and **Teacher's Presentation** PowerPoint slides for whole-class or group display.

National Curriculum

Missing Digit Multiplication covers these Year 5 and Year 6 National Curriculum objectives.

Year 5 Multiplication and division

- Solve problems involving multiplication and division, including using their knowledge of factors and multiples, squares and cubes.
- Multiply numbers up to four digits by a one- or two-digit number using a formal written method, including long multiplication for two-digit numbers.
- Multiply and divide numbers mentally drawing upon known facts.

Year 6 Addition, subtraction, multiplication and division

- Solve problems involving addition, subtraction, multiplication and division.
- Multiply multi-digit numbers up to four digits by a two-digit whole number using the formal written method of long multiplication.

The 6C cycle

The grid on the right shows the relative weighting given to each of the 6 Cs in **Missing Digit Multiplication**.

Curious	★★
Connect	★★
Collaborate	★★★
Communicate	★
Create	★★
Consider	★★

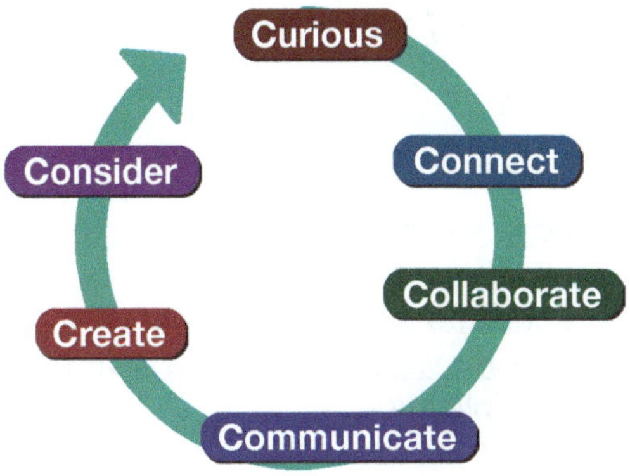

Problem-solving heuristics

The grid shows which problem-solving strategies children are likely to use when investigating the **Connect** tasks in **Missing Digit Multiplication.** These strategies are further exemplified in the **Connect** stage below (pages 74–76).

Draw a diagram	
Make a table / spot a pattern	✓
Make a model / act it out	
Try then improve	✓
Create a systematic list	✓
Solve a simpler but related problem	✓
Work backwards	

Key mathematical ideas

This task requires children to work out what each symbol in the two-digit by two-digit grid calculation represents. It is helpful if children understand the terms partial product, multiplicand and multiplier. Suitable starting points include recognising the position of placeholders and considering which numbers when multiplied by themselves will result in the product having the same digit in the ones column. There are only two options for this: $5 \times 5 = 25$ and $6 \times 6 = 36$. 1×1 would work but does not have a 2 digit-product and so can be discounted. Children are prompted to consider this when solving the simpler but related problem in the **Connect** task. Children should be encouraged to explain why they discount numbers throughout the task.

Answers to the tasks are given on page 78.

Teaching guidance

If you are using the **Express Pathway**, please turn to pages 73–77.

Teaching guidance for the **Full 6C Pathway** is given below.

The **Teacher's Presentation** PowerPoint file has slides specifically mentioned in this teaching guidance. Some of the slides include notes (within the PowerPoint file) providing 'stumble support' suggestions, that may help you to support children in their problem-solving thinking and enable access for children who are less confident.

Awe and wonder

Slide 1

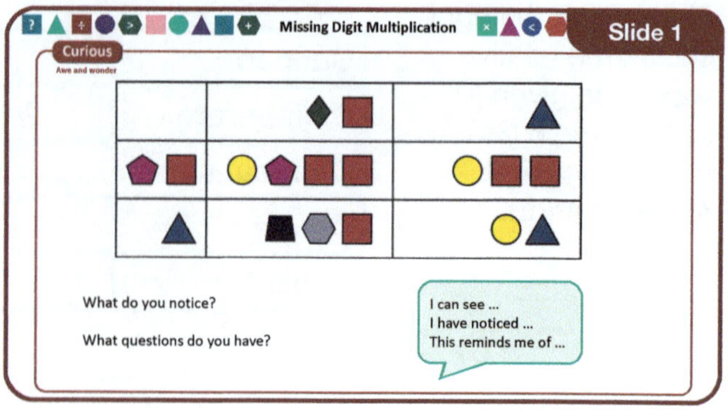

Display slide 1.

- I can see seven different shapes. They are all different colours.
- I can see eight different parts.
- This reminds me of using a grid for multiplication.
- I have noticed some parts of the grid have more / fewer shapes.
- I have noticed some shapes are repeated more than others.
- I have noticed that some of the shapes are in pairs and some are alone.

Decide when to specify that multiplication is the focus.

If children have not used the grid method before, the method could be explored using calculations such as 32 × 4 and 32 × 41 where the multiplication facts should be within children's automatic recall.

Develop and deepen

Slide 2

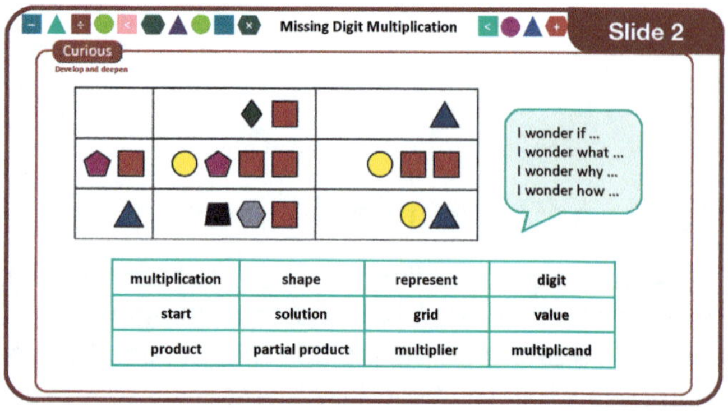

Display slide 2.

Encourage children to use the words in the word bank to develop and deepen their thoughts and questions.

⚠ Children may not have encountered the term 'partial product' previously. If needed, explain that 23 × 24 is made up of the partial products 92 (23 × 4) and 460 (23 × 20). Children may also not have encountered the terms 'multiplier' and 'multiplicand'. Define these words when looking at the word bank so that children can give names to the different parts of the calculation. The multiplicand is the number to be multiplied (23 in the case of 23 × 24). The multiplier is the number by which is it is to be multiplied (24 in this case).

multiplication	**shape**	**represent**	**digit**
start	**solution**	**grid**	**value**
product	**partial product**	**multiplier**	**multiplicand**

Possible lines of enquiry:

- I wonder if we have to add up the shapes.

Possible lines of enquiry linked to the word bank:

- I wonder if we can solve this using multiplication.
- I wonder what a partial product is.
- I wonder what each shape represents.
- I wonder why some shapes occur more than once in the grids.
- I wonder how we can work out what each shape represents.
- I wonder what the best starting point will be.

- I wonder if there is only one solution.
- I wonder how many solutions there are.

Possible lines of enquiry that children may raise now, but should be held back until the **Create** task:

- I wonder if I could create my own puzzle.

Consider whether / how to collate and display all the suggested questions, and whether / how to steer children towards selecting an appropriately challenging question.

Pupil Resources for Full and Express Pathways

Distribute the **Pupil Resources** for **Missing Digit Multiplication** to the children to help support them in their exploration of the different tasks. Use either of the **Curious Pupil Resources** depending on whether you are following the **Full Pathway** or the **Express Pathway**. The **Pupil Resources** for the remaining 5Cs are common to both pathways.

To provide additional flexibility in the classroom, each of the **Pupil Resources** is also provided in the **Teacher's Presentation** PowerPoint file (see slides 3–6), so you can display these at any time if you wish.

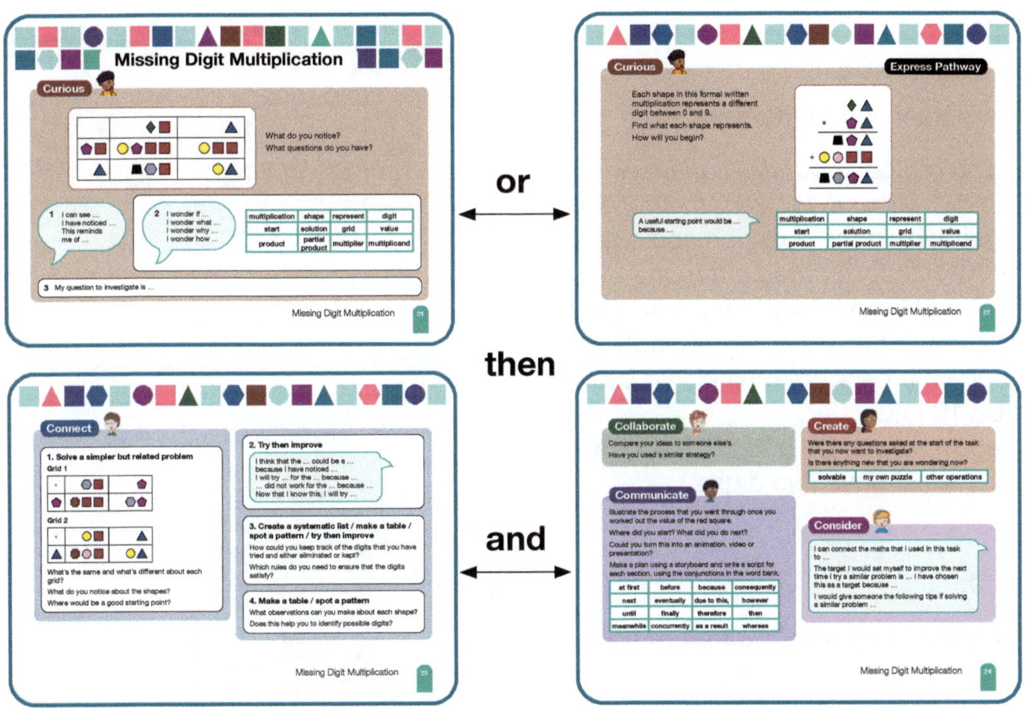

Express Pathway

If you are following the **Express Pathway**, spend a few minutes discussing the **Curious Pupil Resource** with the children. (This is slide 4, so you could display it if you wish.) Use some of the prompts on the previous page to help.

Four tasks involving different strategies are suggested for children to make connections.

1. Solve a simpler but related problem

If children have not calculated using the grid method before or are finding it tricky to find a suitable starting point for the main problem, ask them to consider this simpler problem. You may wish to introduce the grids together, by printing slide 7 and distributing it to children, or separately by displaying slides 8 and 9.

Slide 7

Print slide 7 and distribute to the children.

Ask children to label the columns in the grid with tens and ones so that it is clear to them that, for example, the hexagon and the square in the first row of Grid 1 represent a two-digit number with tens and ones.

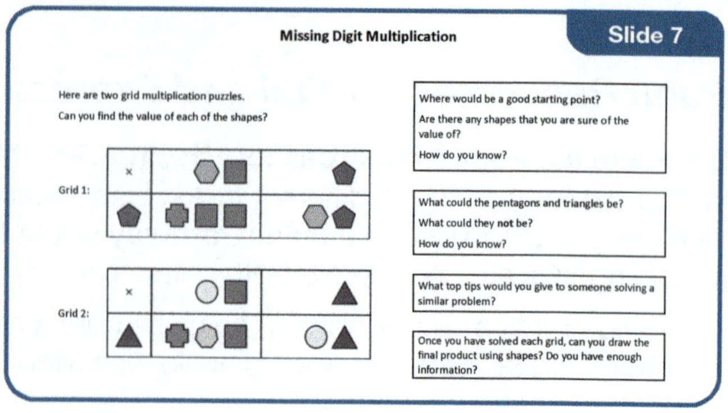

Slide 8

Alternatively, display slide 8.

Draw children's attention to the squares.

- *What do you notice about the squares?*
- *Look at Grid 1. What if we were multiplying by 12? What would the hexagon and square represent? What about the pentagon?*
- The square must represent 0 because the two-digit number has been regrouped into tens and ones. If we were multiplying by 12 it would be 10 and 2.
- The pentagon cannot be a 2. $2 \times 2 = 4$ and the product must have two digits.

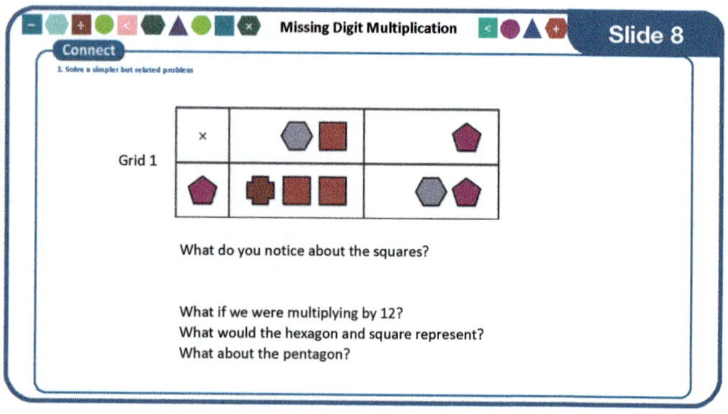

Slide 9

Display slide 9.

- *What's the same and what's different about the grids?*

In these examples, 5 and 6 have been used as the ones digit in the two-digit number as they are the only numbers that when multiplied together to make a two-digit number, end in the same digit. Grid 1 represents 25×5 with the grids reading 100 and 25. Grid 2 represents 36×6 with the grids reading 180 and 36.

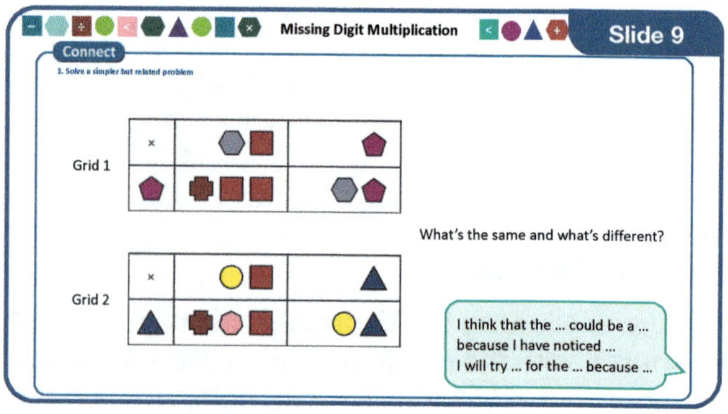

2. Try then improve

Slide 10

Display slide 10.

Children can use try then improve to help them to solve the problem. Encourage them to use the sentence stems on the slide when considering the position of the digits and their reasons for accepting or disregarding their attempts:

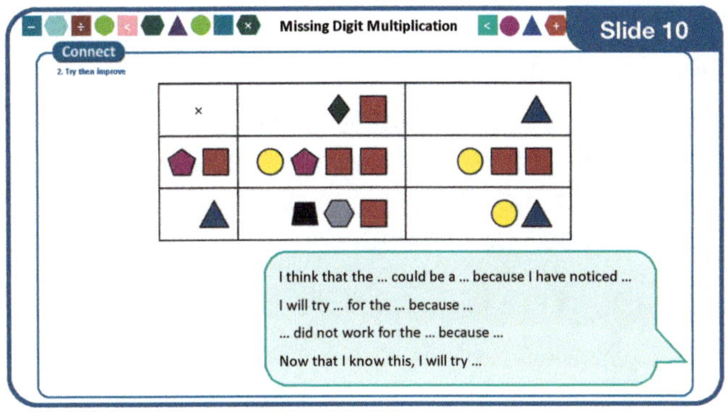

- *Could the triangle be a 4?*
- *Could it be a 5?*

⚠ Watch for children who disregard each arrangement that they make instead of making an improvement each time based on what they have noticed.

3. Create a systematic list / make a table / spot a pattern / try then improve

Slide 11

Display slide 11.

Once the rules needed to satisfy the value of the triangle have been ascertained (triangle × triangle must give a two-digit product; the ones digit in the product must be the same as the multiplier and the multiplicand), children could begin by testing which digits would work and which they can eliminate. These could be recorded in a table such as the one detailed on the slide.

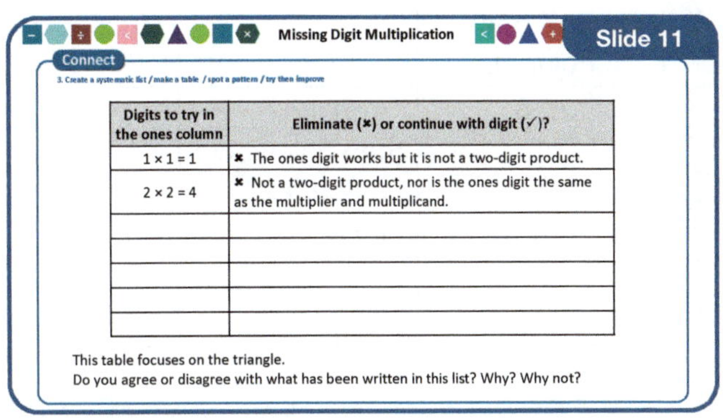

Children may need support with how to organise their table into a systematic order. Some children may eliminate some of the digits straight away using their knowledge of two-digit products and so start with 4 × 4.

Can they articulate the choices that they have made?

Digits to try in the ones column	Eliminate (✗) or continue with digit (✓)?
1 × 1 = 1	✗ The ones digit works but it is not a two-digit product.
2 × 2 = 4	✗ Not a two-digit product nor is the ones digit the same as the multiplier and multiplicand.
3 × 3 = 9	✗ As above.
4 × 4 = 16	✗ Two-digit product but the ones digit is not the same as the multiplier and multiplicand.
5 × 5 = 25	✓ Satisfies both – will continue to work with.
6 × 6 = 36	✓ Satisfies both – will continue to work with.
7 × 7 = 49	✗ Two-digit product but the ones digit is not the same as the multiplier and multiplicand.

Completed table for the triangle

Connect

4. Make a table / spot a pattern

Slide 12

Print slide 12 and distribute to children.

Ask children to use the table to track their observations and show how these observations lead to decisions about which digits to try and then which can be eliminated.

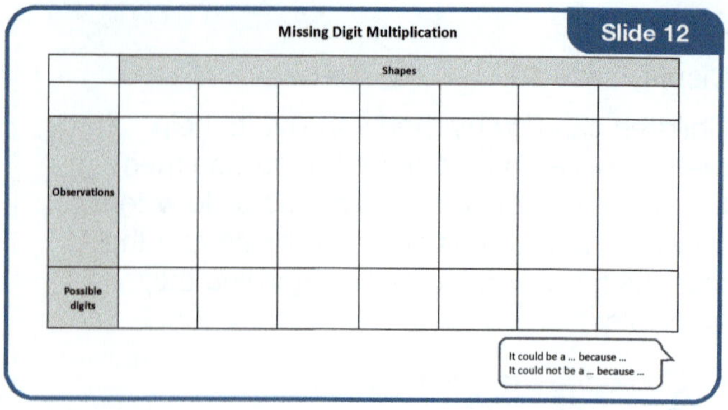

	Shapes						
	▲	●	■	⬟	⬢	▬	◆
Observations	one-digit × one-digit = two-digit Product has the same digit in the ones column.	Appears multiple times in partial products.	Appears multiple times. Must be placeholder.	Appears twice.	Appears once in a partial product.	Appears once in a partial product.	Appears once.
Possible digits	5, 6	If triangle = 5 then this would be a 2. If triangle = 6 then this would be a 3.	0	If ●■■ = 200 then this would be a 4. If ●■■ = 300 then this would be a 5.			

Example completed table

- The triangle can't be a 5 because this would make the pentagon a 4. It can't be a 4 because to make 2400 the diamond would have to be a 6, and this does not work with the bottom number in that column (which doesn't contain 00).

Collaborate

Refer to the **Collaborate** task of the **Pupil Resource**.

Ask children to work in pairs, either grouping them according to their preferred strategies or on a more random basis.

- *Compare your ideas to someone else.*
- *Have you used a similar strategy?*

Children could then work in pairs, or small groups to create their film or animation as detailed in the **Communicate** task.

Refer to the **Communicate** task of the **Pupil Resource**.

Slide 13

Print slide 13 and distribute to children.

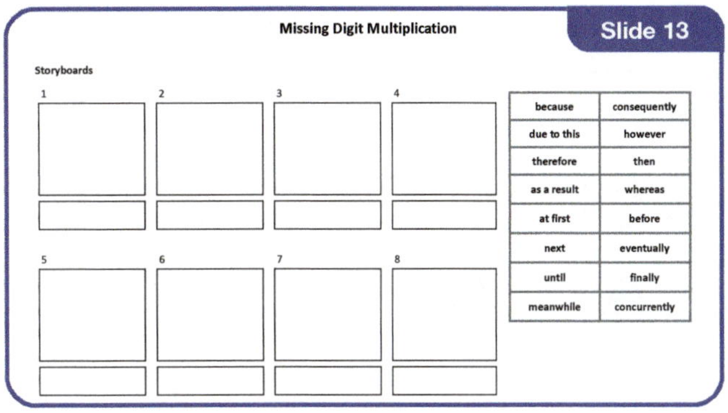

Children could present their process visually, showing where they started and where they went next. Stop-motion animation, a short video or a live demonstration could be used to show how, once it is ascertained that the square represents 0, the rest of the puzzle was solved.

Children can use the storyboards to sequence the steps they took and write a script to describe the reasoning behind their decisions.

Refer to the **Create** task of the **Pupil Resource**.

Slide 14

Display slide 14.

- *Were there any questions asked at the start of the task that you now want to investigate?*
- *Is there anything new that you are wondering now?*

Children can use the word bank to prompt further questions of their own.

solvable	my own puzzle	other operations

- I wonder if I could create my own puzzle.
- *What have you learned from solving the puzzle that will help you make your own?*
- *How will you make sure that it is solvable?*

Refer to the **Consider** task of the **Pupil Resource**.

Slide 15

Display slide 15.

Encourage children to think about the strategies that they have used. The prompts below may help get children started.

Cognitive learning domain

- The strategies that helped me in this task were …
- I can connect the maths that I used in this task to …
- I might be able to use … in the future because …

Affective domain

- The fun bit of my learning was … It was fun because …
- I said 'aha' when … I said it because …
- I was surprised about … because … This led me to think about …

Metacognitive domain

- The target I would set myself to improve the next time I try a similar problem is … I have chosen this as a target because …
- I would give someone the following tips if solving a similar problem …

Answers

The partial products are 36, 300, 420 and 3500, which combine to make the product 4256.

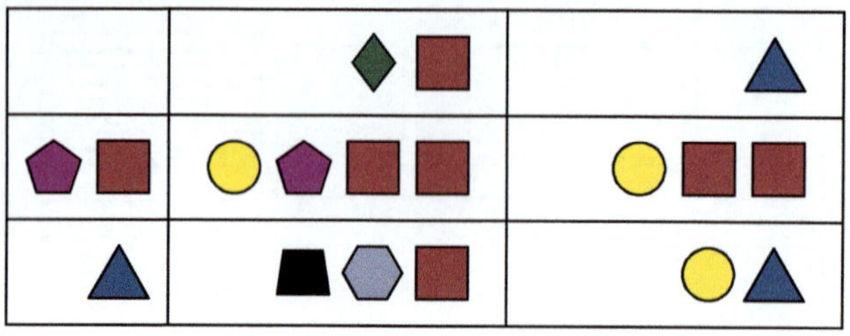

×	70	6
50	3500	300
6	420	36

Express Pathway

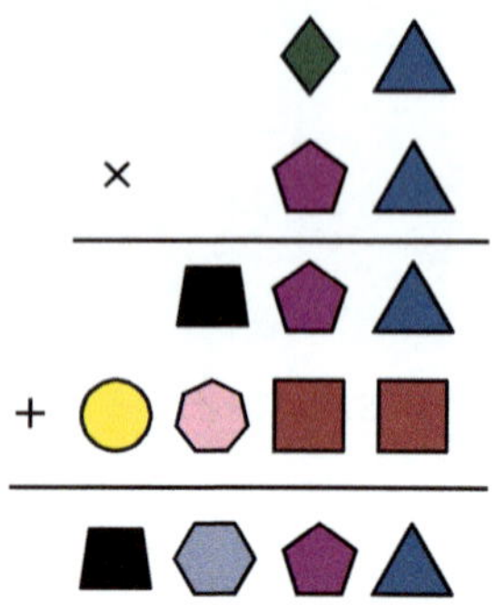

$$\begin{array}{r} {}^{3}\;\,{}^{3}\;\;\;\;\; \\ 7\;6 \\ \times\;\;\;5\;6 \\ \hline 4\;5\;6 \\ +\;3\;8\;0\;0 \\ \hline 4\;2\;5\;6 \\ \hline {}_{1}\;\;\;\;\;\;\;\;\; \end{array}$$

Simpler but related problem

Grid 1

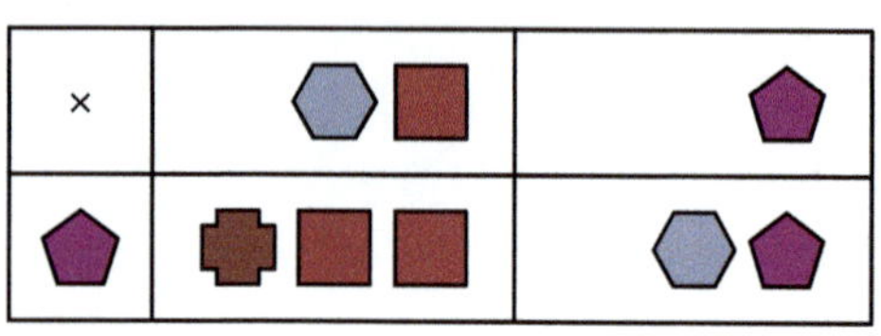

×	20	5
5	100	25

Grid 2

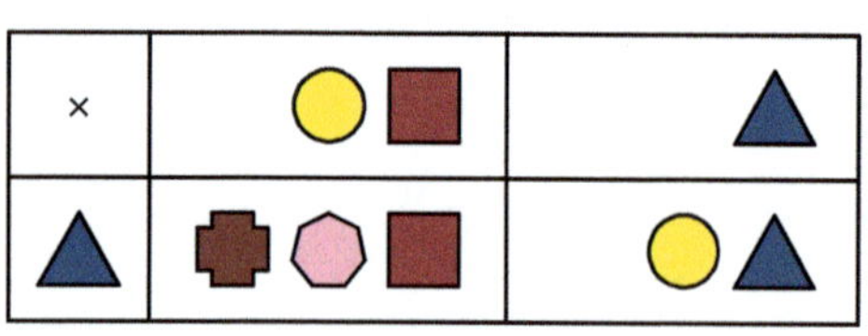

×	30	6
6	180	36

In all the problems in this task, each symbol represents a single digit:

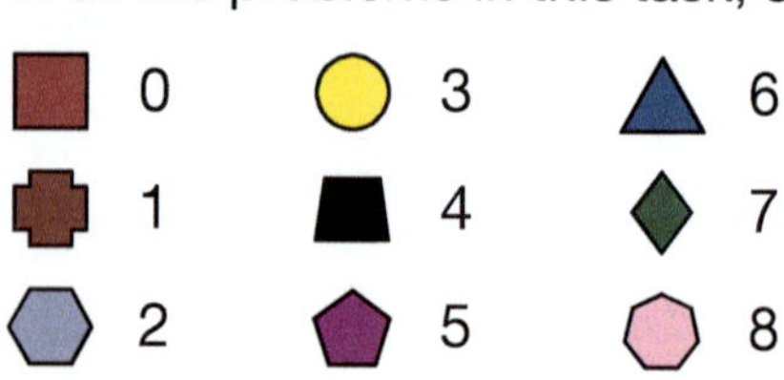

The Perfect Number

Year 6

Introduction

The Perfect Number gives children an opportunity to revisit and rehearse their knowledge of factors, composite (non-prime) and prime numbers and to investigate the properties of perfect numbers.

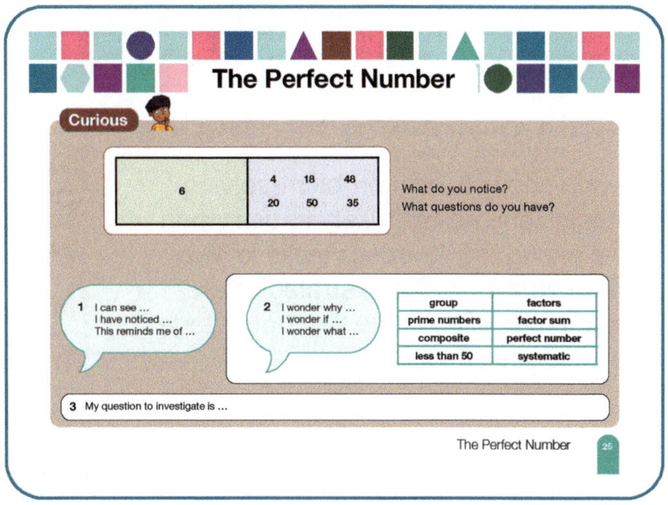

Children discuss the properties of some given numbers and why they may have been sorted in a certain way. After children have been introduced to the properties of a perfect number, their task is to find other perfect numbers within a certain range. They decide which numbers to rule out immediately and which to test further, working efficiently and systematically. Children practise working as a group, underlining the importance of explaining their reasoning as well as considering and valuing the suggestions of others.

This task promotes greater-depth learning in a range of ways, supporting children to identify and articulate underlying mathematical structures. For example, in the **Connect** stage, greater depth is promoted by asking children to explore a range of different mathematical resources and models, choose appropriate resources for the task, and explain their choice. They will also monitor their progress towards the solution, and adjust their strategy if needed.

This **Teacher Guide** is supported by photocopiable **Pupil Resources** and **Teacher's Presentation** PowerPoint slides for whole-class or group display.

National Curriculum

The Perfect Number covers these Year 6 National Curriculum objectives.

Addition, subtraction, multiplication and division

- Identify common factors, common multiples and prime numbers.
- *Pupils continue to use all the multiplication tables to calculate mathematical statements in order to maintain their fluency.* [non-statutory guidance]

The 6C cycle

The grid on the right shows the relative weighting given to each of the 6 Cs in **The Perfect Number**.

Curious	★★
Connect	★★★
Collaborate	★★★
Communicate	★★
Create	★★
Consider	★★

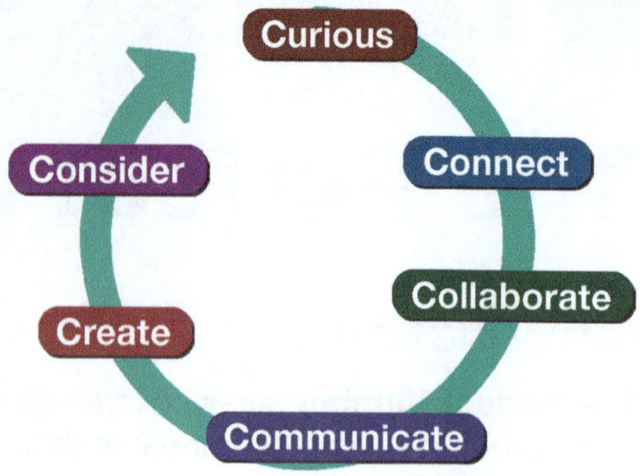

Problem-solving heuristics

The grid shows which problem-solving strategies children are likely to use when investigating the Connect tasks in **The Perfect Number**. These strategies are further exemplified in the Connect stage below (pages 83–85).

Draw a diagram	
Make a table / spot a pattern	✓
Make a model / act it out	✓
Try then improve	
Create a systematic list	✓
Solve a simpler but related problem	✓
Work backwards	

Key mathematical ideas

Perfect numbers are numbers where the sum of its factors (excluding the number itself) is equal to the number. 6 and 28 are the only perfect numbers less than 100.

$6 = 1 + 2 + 3$ $\quad$ $28 = 1 + 2 + 4 + 7 + 14$

The next two perfect numbers are 496 and 8128.

$496 = 1 + 2 + 4 + 8 + 16 + 31 + 62 + 124 + 248$

$8128 = 1 + 2 + 4 + 8 + 16 + 32 + 64 + 127 + 254 + 508 + 1016 + 2032 + 4064$

This investigation will also lead children to discover numbers where the factor sum (excluding the number itself) is greater than the number. These are called abundant numbers. For example, 20 and 100 are abundant numbers.

$20 < 1 + 2 + 4 + 5 + 10$ $\quad$ $100 < 1 + 2 + 4 + 5 + 10 + 20 + 25 + 50$

Numbers where the factor sum (excluding the number itself) is less than the number are called deficient numbers. For example, 9 and 14:

$9 > 1 + 3$ $\quad$ $14 > 1 + 2 + 7$

Prime numbers are always deficient numbers.

Answers to the tasks are given on page 87.

Teaching guidance

If you are using the **Express Pathway**, please turn to pages 82–87.

Teaching guidance for the **Full 6C Pathway** is given below.

The **Teacher's Presentation** PowerPoint file has slides specifically mentioned in this teaching guidance. Some of the slides include notes (within the PowerPoint file) providing 'stumble support' suggestions, that may help you to support children in their problem-solving thinking and enable access for children who are less confident.

Awe and wonder

Slide 1

Display slide 1.

Allow children to share what they are noticing or wondering.

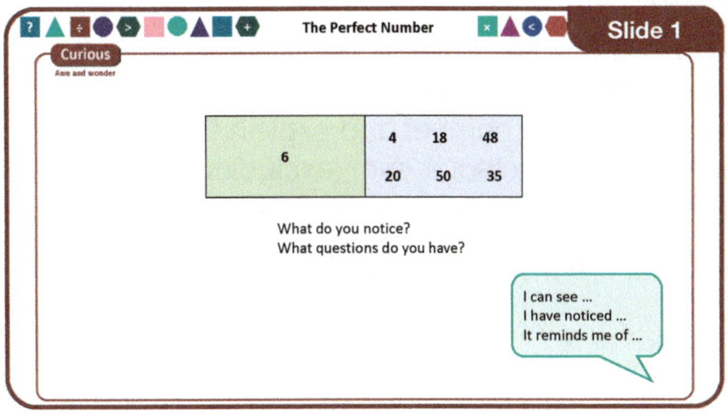

- I can see two groups of numbers.
- I have noticed that 6 is on its own and there are 6 other numbers in the other group.
- I can see that some of the larger group are multiples of 6 but some aren't.
- I have noticed that one of the numbers in the larger group is less than 6.
- I have noticed that there is only one odd number in either group.
- It reminds me of a sorting diagram.

Develop and deepen

Slide 2

Display slide 2.

Encourage children to use the words in the word bank to develop and deepen their thoughts and questions.

⚠ Children may not have encountered the term 'factor sum' before and may benefit from revisiting 'digit sum'.

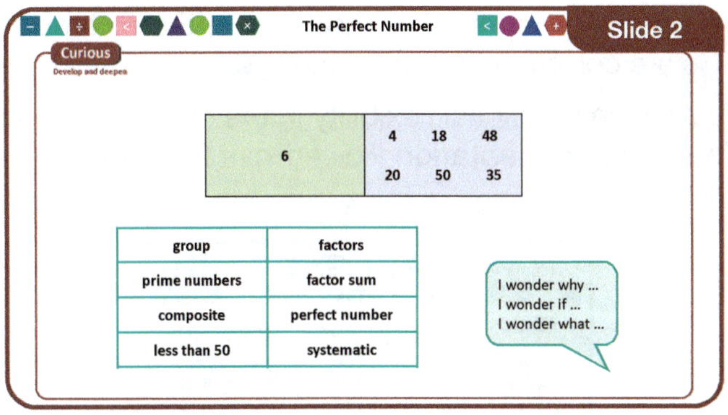

group	**factors**
composite	**perfect number**
prime numbers	**factor sum**
less than 50	**systematic**

Possible lines of enquiry:

- I wonder why 6 is on its own.
- I wonder if there are any other numbers that would go in the same group as 6.

Possible lines of enquiry linked to the word bank:

- I wonder if the grouping is to do with factors and composite numbers.
- I wonder if any prime numbers would go in either group.
- I wonder what the factor sum of each number is.
- I wonder what a perfect number is.
- I wonder if 6 is a perfect number.
- I wonder if I could find other perfect numbers by working systematically.

Possible lines of enquiry that children may raise now, but should be held back until the **Create** task:

- I wonder if there are any perfect numbers over 50.
- I wonder if there are numbers whose factor sum is larger than the number or smaller than the number.

Consider whether / how to collate and display all the suggested questions, and whether / how to steer children towards selecting an appropriately challenging question.

Slide 3

Display slide 3.

Now that children have developed their questions with the word bank, reveal that 6 is called a perfect number and explain that perfect numbers have a factor sum (excluding the number itself) which is equal to the number.

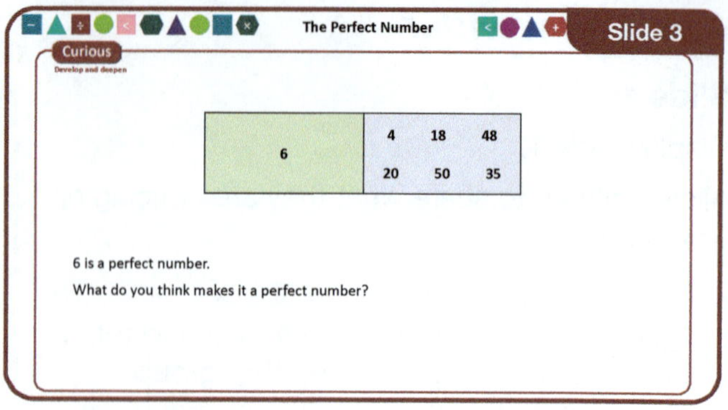

- *What are the factors of 6?*

If necessary, remind children that the factors of a number divide that number exactly.

- *The factors of 6 are 1, 2, 3 and 6.*
- *So the factor sum for 6, excluding the number 6 itself = 1 + 2 + 3 = 6.*

Pupil Resources for Full and Express Pathways

Distribute the **Pupil Resources** for **The Perfect Number** to the children to help support them in their exploration of the different tasks. Use either of the **Curious Pupil Resources** depending on whether you are following the **Full Pathway** or the **Express Pathway**. The **Pupil Resources** for the remaining 5Cs are common to both pathways.

To provide additional flexibility in the classroom, each of the **Pupil Resources** is also provided in the **Teacher's Presentation** PowerPoint file (see slides 4–7), so you can display these at any time if you wish.

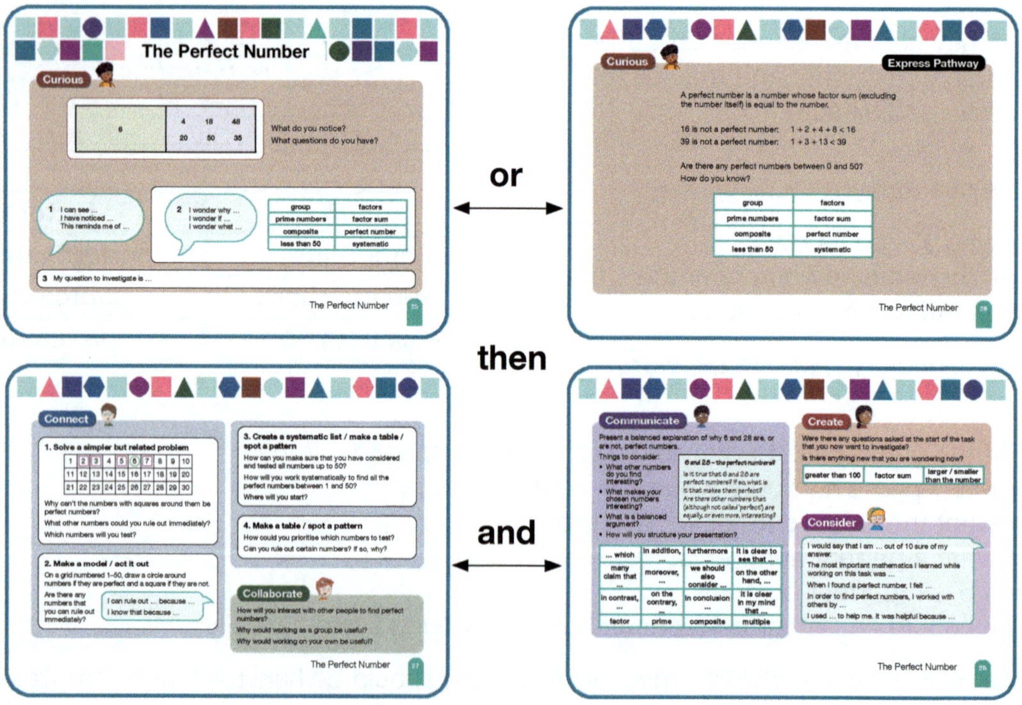

Express Pathway

If you are following the **Express Pathway**, spend a few minutes discussing the **Curious Pupil Resource** with the children. (This is slide 5, so you could display it if you wish.) Use some of the prompts on the previous page to help.

Four tasks involving different strategies are suggested for children to make connections.

1. Solve a simpler but related problem

Slide 8

Display slide 8 which shows a smaller range of numbers for children to test to see if they are perfect numbers.

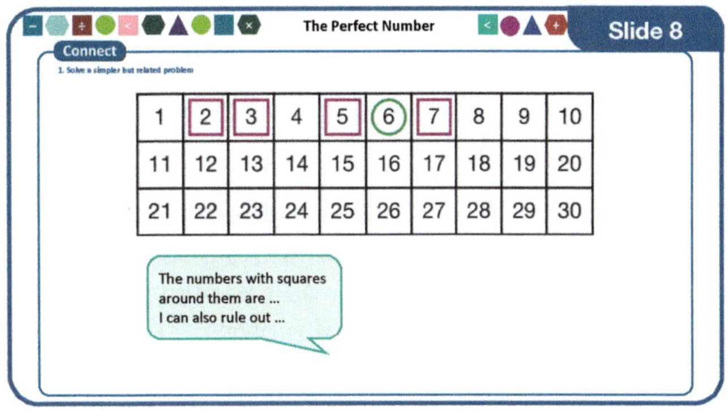

- *Why is there a circle around number 6?*
- We already know that 6 is a perfect number.
- *Tell me about the numbers that have squares around them. Do you think they are perfect numbers?*

Children should spot that the numbers with squares around them are prime numbers.

- *What do you know about the factors of prime numbers? How do we know that their factor sums will not be equal to the number? (Remember to disregard the number itself.)*
- 5 has 1 and 5 as factors. To find a perfect number I need to add the factors but not the number itself. This means that the factor sum would be 1. So 5 is not a perfect number. I think all prime numbers will have 1 as their factor sum if we don't include the number itself.

Slide 9

Display slide 9 which shows the numbers 1–30 with all the prime numbers indicated.

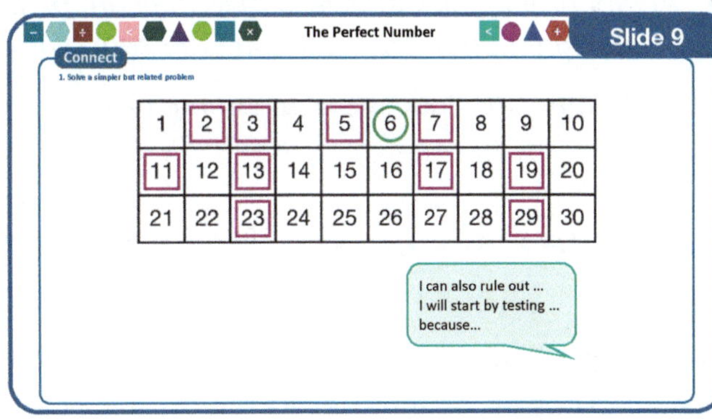

- *Which other numbers shall we test? Where could you start?*

2. Make a model / act it out

Additional resources required:
1–50 number grids

Slide 10

Display slide 10 and give children 1–50 number grids to use in their investigation.

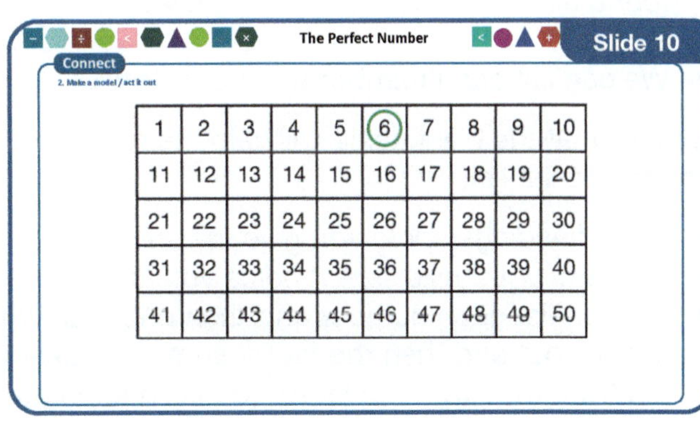

- *Are there any numbers that you can eliminate immediately? Why?*
- I know that all prime numbers only have 1 and themselves as factors. We don't include the number itself in the factor sum, therefore all prime numbers have a factor sum of 1, so none are equal to the number. A prime number can't be a perfect number.

⚠ Look out for children who are not secure with their understanding of prime numbers or who include the number itself in the factor sum, for example, $3 < 1 + 3$.

Slide 11

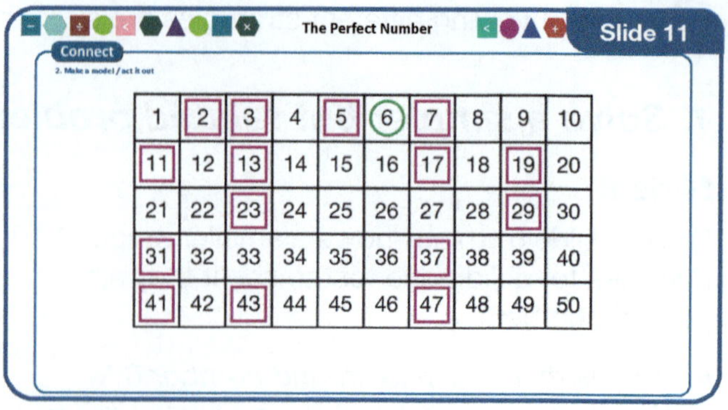

Display slide 11 which shows the same 1–50 grid but with all the prime numbers highlighted.

Encourage children to use their prior knowledge of factors. Ask them to consider numbers that they know have many or few factors and reason whether this makes them more or less likely to be a perfect number.

- I know that 9 will not be a perfect number as it only has factors of 1 and 3 and so will not equal 9. 49 is similar as the only factors to total are 1 and 7.
- I know that 36 has lots of factors – 1, 2, 3, 4, 6, 9, 12 and 18 and so the factor sum will be greater than 36. 12, 24, 30 and 48 have a number of factors too so I don't think that they will be perfect numbers. I know that the factor sums will exceed the number so will not test them.
- *Is this the same for all square numbers? Do any square numbers have lots of factors?*
- I think that square numbers that are even numbers, like 64 and 100, will have lots of factors. 64 has factors 1, 2, 4, 8, 16 and 32, not including itself. 100 also has lots of factors: 1, 2, 4, 5, 10, 20, 25, 50.

Once children have found numbers where the factor sum exceeds the number, you may choose to introduce children to the fact that these are called abundant numbers.

- *An abundant number is less than the sum of its factors, excluding the number itself.*
- *A deficient number is more than the sum of its factors, excluding the number itself.*

Children may conclude that if a number has lots of factors, then it is likely to be abundant.

3. Create a systematic list / make a table / spot a pattern

Slide 12

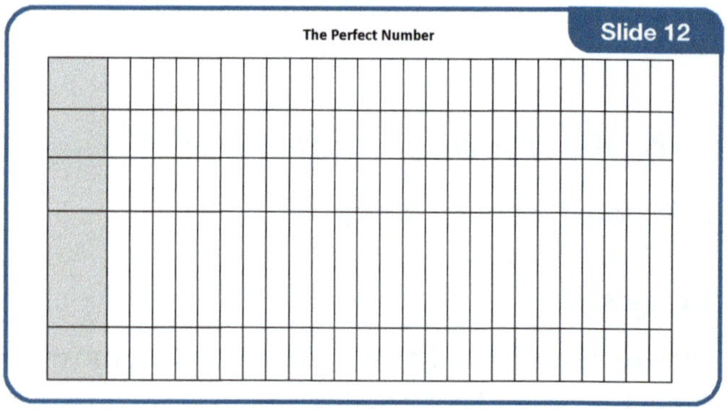

Print slide 12 and distribute to children (they will need two copies, so they have a row for each of the numbers 1–50).

- *How can you be systematic in your testing so that you can be confident you have found all the perfect numbers between 1 and 50?*
- We can list each number and its factors.

Explain that the blank table will help them to structure their list.

- *What headings could you give the table?*
- *How will you make sure it helps you test the numbers correctly to see if they are perfect?*
- We can list each number and then prioritise which numbers to test first. We should write the factors out and then the factor sum. To make sure we don't include the number itself we could subtract that from the factor total in a separate column.

Agree the headings as follows and ask children to fill them in their table. Then ask them to write the numbers 1–50 in the 'Number' column. They can then prioritise which numbers they want to test first, and complete the rows for those numbers.

- *Where will you start from?*

Number	Factors	Factor sum	Factor sum minus number itself	Perfect number?

4. Make a table / spot a pattern

Slide 13

Print slide 13 and distribute to children.

Encourage them to consider which numbers should be prioritised for testing and complete the table with their reasoning. This will depend on their prior experience with factors.

They can use the number grid on the slide to mark off the numbers as they sort them into their chosen group.

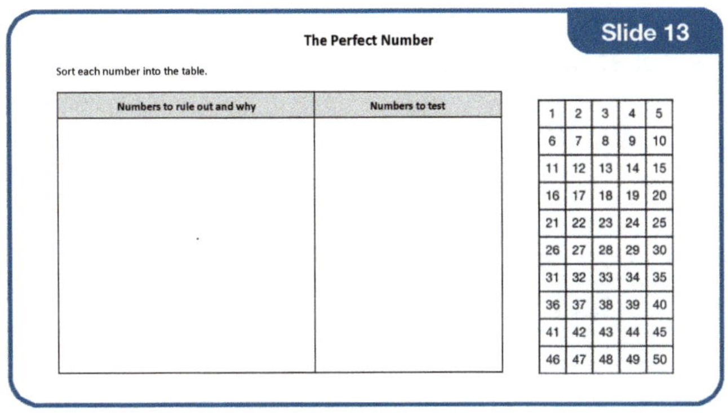

Numbers to rule out and why	Numbers to test
1 – only itself as a factor *2, 3, 5, 7, 11, 13, 17, 19, 23, 29, 31, 37, 41, 43, 47 – all prime numbers so only 1 as a factor* *4, 8, 10, 14, 22 – factors too low to equal number* *9, 25, 49 – I know that odd square numbers have few factors* *6 – I know now that this is a perfect number* *1 + 2 + 3 = 6*	*12, 15, 16, 18, 20, 21, 26, 27, 28, 32, 33, 34, 35, 38, 39, 42, 44, 45, 46, 50* *24, 30, 36, 40, 48 – I predict that these factor sums will be greater than the numbers as they have lots of factors!*

Example completed table

Ask children to discuss their choices with a partner.

Refer to the **Collaborate** task of the **Pupil Resource**.

This task would work well with children exploring different number ranges in small groups. For example, one group member could be testing numbers between 10 and 20 while another looks at numbers between 20 and 30, and so on. They can then share their findings within their group, before comparing with other groups as well.

Children could then work in pairs to create their explanation text or presentation as detailed in the **Communicate** task.

Communicate

Refer to the Communicate task of the **Pupil Resource**.

Slide 14

Display slide 14.

This task provides an opportunity for the children to present a balanced explanation about why 6 and 28 are or are not perfect numbers.

Children can choose between a written or oral presentation.

The sample opening text and word bank could be useful to help children to structure and demonstrate their reasoning.

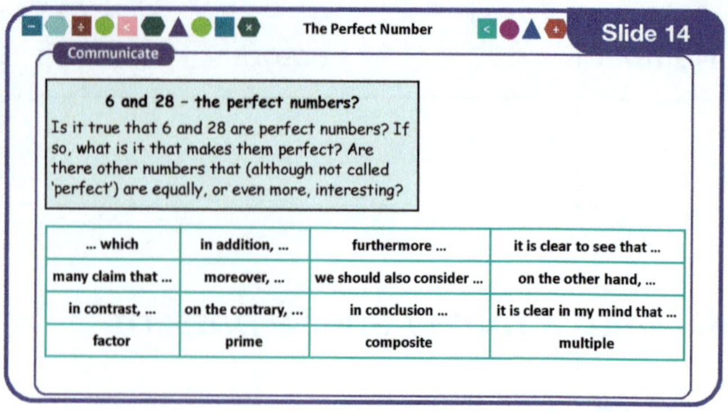

... which	**in addition, ...**	**furthermore ...**	**it is clear to see that ...**
many claim that ...	**moreover, ...**	**we should also consider ...**	**on the other hand, ...**
in contrast, ...	**on the contrary, ...**	**in conclusion ...**	**it is clear in my mind that ...**
factor	**prime**	**composite**	**multiple**

Create

Refer to the Create task of the **Pupil Resource**.

Slide 15

Display slide 15.

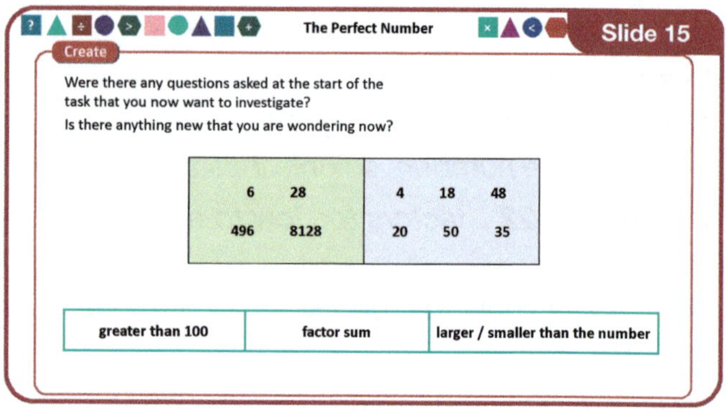

- *Were there any questions asked at the start of the task that you now want to investigate?*
- *Is there anything new that you are wondering now?*

Children can use the word bank to prompt further questions of their own.

greater than 100	**factor sum**	**larger / smaller than the number**

- I wonder if there are any perfect numbers over 50.
- I wonder if there are numbers whose factor sum is larger than the number or smaller than the number.
- I wonder if there are any perfect numbers greater than 100.

Refer to the **Consider** task of the **Pupil Resource**.

Encourage children to think about the strategies that they have used. The prompts below may help get children started.

Cognitive learning domain

- I would say that I am … out of 10 sure of my answer.
- The most important mathematics I learned while working on this task was …
- In the future I would use what I have found out by …

Affective domain

- The fun bit of my learning was … It was fun because …
- When I found a perfect number, I felt …
- I would say that I am … out of 10 pleased with my answer.

Metacognitive domain

- In order to find perfect numbers, I worked with others by …
- I used … to help me. It was helpful because …

Answers

6 and 28 are the only perfect numbers less than 100.

The next two perfect numbers are 496 and 8128.

In the 1–50 number grid below:

- perfect numbers are indicated by green circles;
- abundant numbers are indicated by blue pentagons;
- deficient numbers are indicated by purple squares (with prime numbers indicated by a purple dot).

1	2	3	4	5	6	7	8	9	10
11	12	13	14	15	16	17	18	19	20
21	22	23	24	25	26	27	28	29	30
31	32	33	34	35	36	37	38	39	40
41	42	43	44	45	46	47	48	49	50

Algebra is Amazing

Year 6

Introduction

Algebra is Amazing provides children with the excitement of performing a trick alongside the opportunity to develop their understanding of algebra. Children explore the concept primarily with models and tables. These two heuristics are closely linked in the task and work together to support children to develop a deep conceptual understanding of algebraic notation.

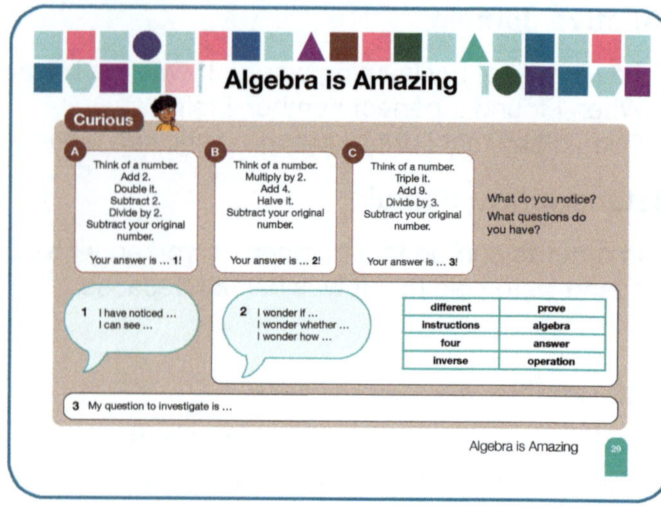

Children then collaborate to expand their range of 'tricks', manipulating the mathematical language they use to disguise the sequences in order to stun their audience. It must be remembered that performing can cause anxiety for some children. Therefore other opportunities, such as digital delivery, should be considered.

This task promotes greater-depth learning in a range of ways, supporting children to identify and articulate underlying mathematical structures. For example, in the **Connect** stage, greater depth is promoted by asking children to explore a range of different mathematical resources and models, choose appropriate resources for the task, and explain their choice. They will also monitor their progress towards the solution, and adjust their strategy if needed.

This **Teacher Guide** is supported by photocopiable **Pupil Resources** and **Teacher's Presentation** PowerPoint slides for whole-class or group display.

National Curriculum

Algebra is Amazing covers these Year 6 National Curriculum objectives.

Algebra

- Express missing number problems algebraically.

Addition, subtraction, multiplication and division

- Use their knowledge of the order of operations to carry out calculations involving the four operations.

The 6C cycle

The grid on the right shows the relative weighting given to each of the 6 Cs in **Algebra is Amazing.**

Curious	★★
Connect	★★★
Collaborate	★★
Communicate	★
Create	★★
Consider	★★

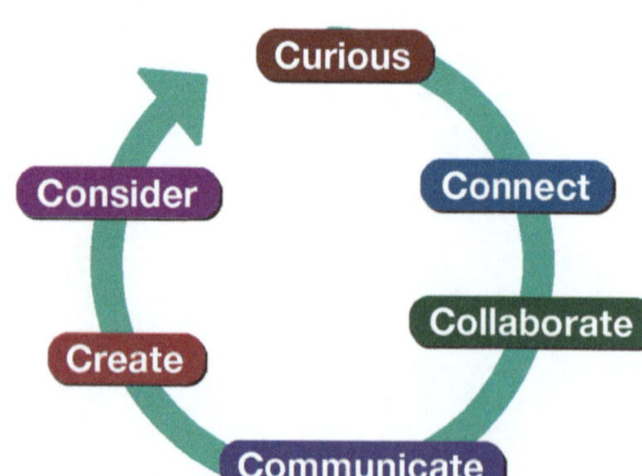

Problem-solving heuristics

The grid shows which problem-solving strategies children are likely to use when investigating the **Connect** tasks in **Algebra is Amazing**. These strategies are further exemplified in the **Connect** stage below (pages 92–94).

Draw a diagram	
Make a table / spot a pattern	✓
Make a model / act it out	✓
Try then improve	
Create a systematic list	
Solve a simpler but related problem	✓
Work backwards	

Key mathematical ideas

Algebra is often intimidating, yet it does not need to be. In this task, children start to explore one aspect of algebra: that an algebraic symbol can represent a quantity without a fixed value, commonly known as a variable. **Algebra is Amazing** allows children to explore algebraic patterns with a fun and motivating task.

Children do not need to express their solutions algebraically, as they can explain their solutions pictorially or through written descriptions. However, this task does provide an opportunity to show children the power of algebra and allows them to explore how algebra can be used to express patterns. These patterns can be seen clearly by building each stage of the calculation using number rods and then comparing to the algebraic form. Children can use any length of number rod to represent their starting number, except for the 1-rod. For example:

Number rod to represent the number you thought of when asked to think of a number.

Variable that represents the number you thought of when asked to think of a number.

Think of a number.	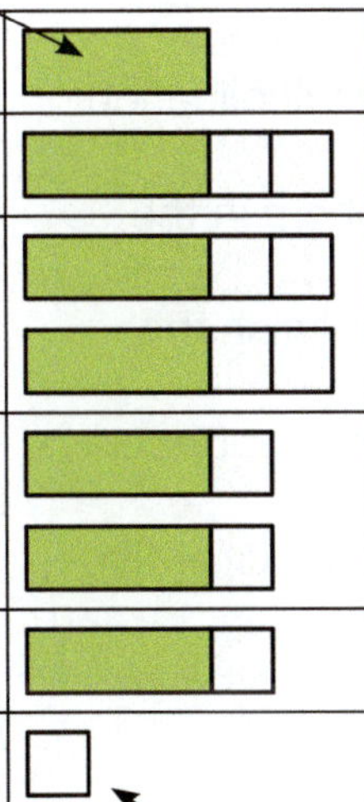	n
Add 2.		$n+2$
Double it.		$2(n+2)=2n+4$
Subtract 2.		$2n+2$
Divide by 2.		$n+1$
Subtract your original number.		1

Your answer

Answers to the tasks are given on page 96.

Teaching guidance

If you are using the **Express Pathway**, please turn to pages 91–96.

Teaching guidance for the **Full 6C Pathway** is given below.

The **Teacher's Presentation** PowerPoint file has slides specifically mentioned in this teaching guidance. Some of the slides include notes (within the PowerPoint file) providing 'stumble support' suggestions, that may help you to support children in their problem-solving thinking and enable access for children who are less confident.

Awe and wonder

Reveal slides 1–3 one at a time, allowing time for discussion and observations as prompted below.

Tell children that you are going to amaze them by predicting the number that they think of and therefore the answer to a mathematical sequence.

You may wish to up the anticipation by placing the answers in envelopes and giving them to children to hold for the big reveal.

Slides 1, 2 and 3

Display slide 1 and read through each step of the sequence with the children. Reveal that you know they all have the answer 1.

Repeat for slides 2 and 3.

⚠ Some children may benefit from having access to resources to make sure they complete the calculation accurately.

- I have noticed that you always get it right.
- I have noticed that all the instructions involve addition, subtraction, multiplication and division.
- I have noticed that all the numbers in the instructions are small.
- I can see that they all start and finish the same way.
- I can see that they all finish with subtract your original number.
- I have noticed that they all have four or five different steps.

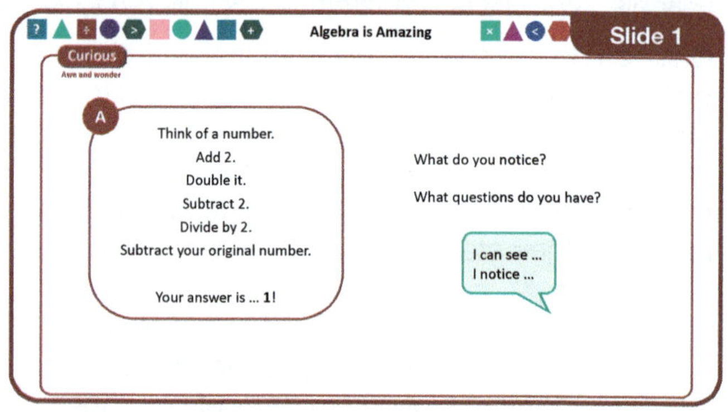

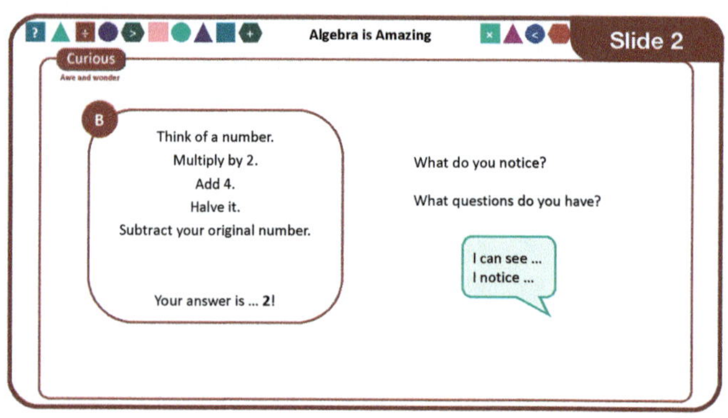

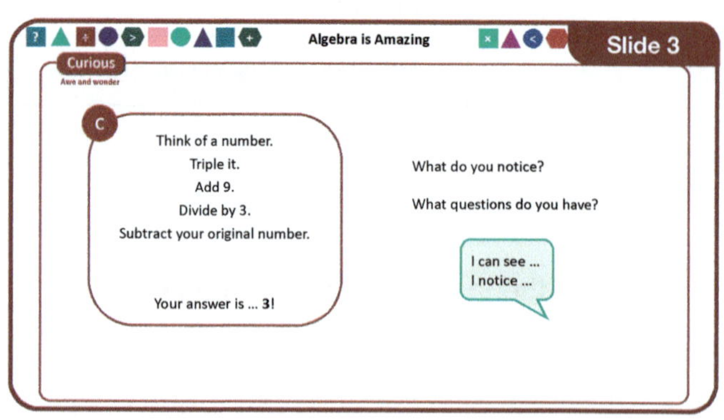

Develop and deepen

Encourage children to use the words in the word bank to develop and deepen their thoughts and questions.

different	**instructions**	**four**	**inverse**
prove	**algebra**	**answer**	**operation**

Possible lines of enquiry linked to the word bank:

- I wonder if I can make a set of instructions that have the answer 4.
- I wonder whether I can write a different set of instructions with the answer 1, 2 or 3.
- I wonder how I can prove how these 'tricks' work.
- I wonder if I can use algebra.

Possible lines of enquiry that children may raise now, but should be held back until the **Create** task:

- I wonder if there is a set of instructions which can be changed to make any answer possible.

Consider whether / how to collate and display all the suggested questions, and whether / how to steer children towards selecting an appropriately challenging question.

Pupil Resources for Full and Express Pathways

Distribute the **Pupil Resources** for **Algebra is Amazing** to the children to help support them in their exploration of the different tasks. Use either of the **Curious Pupil Resources** depending on whether you are following the **Full Pathway** or the **Express Pathway**. The **Pupil Resources** for the remaining 5Cs are common to both pathways.

To provide additional flexibility in the classroom, each of the **Pupil Resources** is also provided in the **Teacher's Presentation** PowerPoint file (see slides 4–7), so you can display these at any time if you wish.

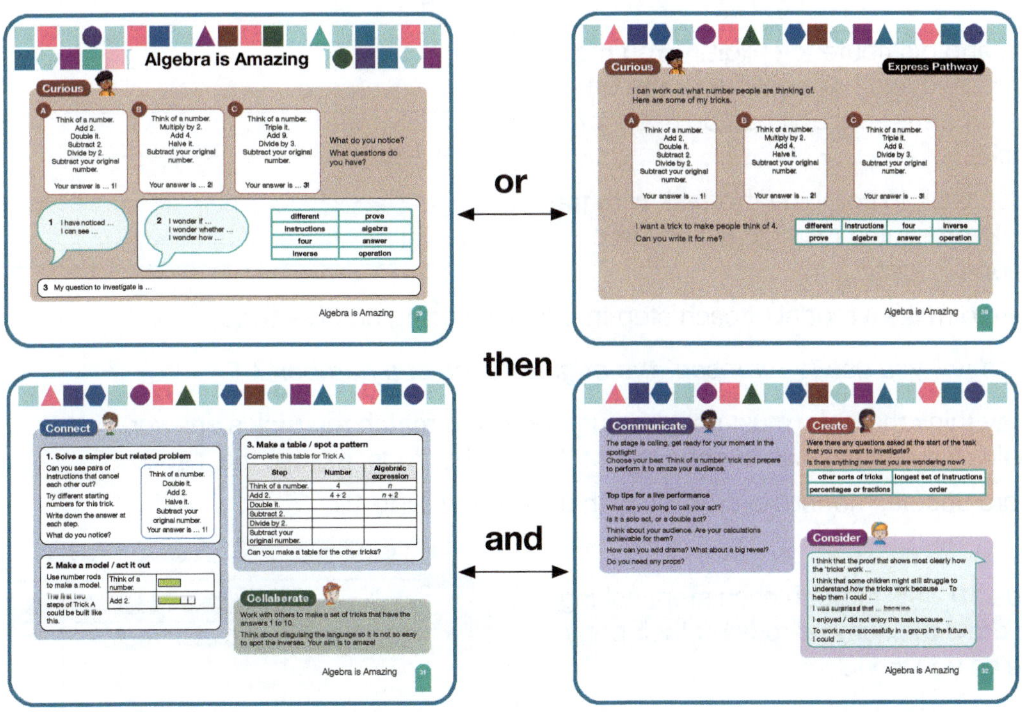

Express Pathway

If you are following the **Express Pathway**, spend a few minutes discussing the **Curious Pupil Resource** with the children. (This is slide 5, so you could display it if you wish.) Use some of the prompts on the previous page to help.

Three tasks involving different strategies are suggested for children to make connections.

1. Solve a simpler but related problem

Slide 8

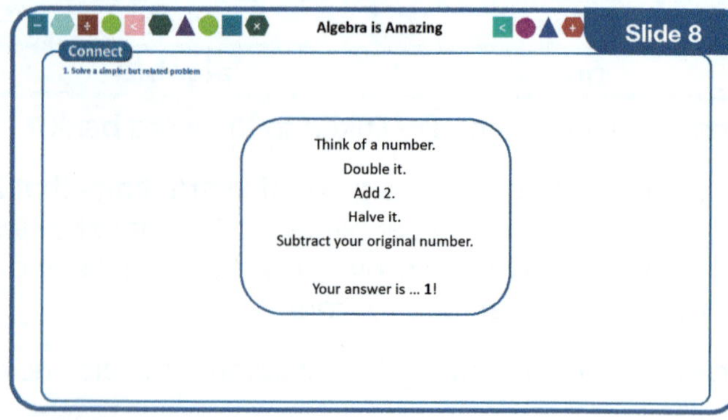

Display slide 8, which presents children with a slightly simpler problem where the pairs of inverse operations are easier to spot. Let them explore with different starting numbers, encouraging them to write down the answer to each step of the problem.

For each calculation draw children's attention to the inverse operations.

Ask them to look at the instructions 'Think of a number' and 'Subtract your original number', and 'Double it' and Halve it'.

- *What do these pairs of instructions do?* (They cancel each other out.)
- *Why do you think we add 2?*
- The answer needs to be 1 and we are going to halve it in the next step (2 halved is 1).
- *What would you add if you want the answer to be 5?*
- Double 5 is 10 so I would add 10.

2. Make a model / act it out

Additional resources required: number rods

Slide 9

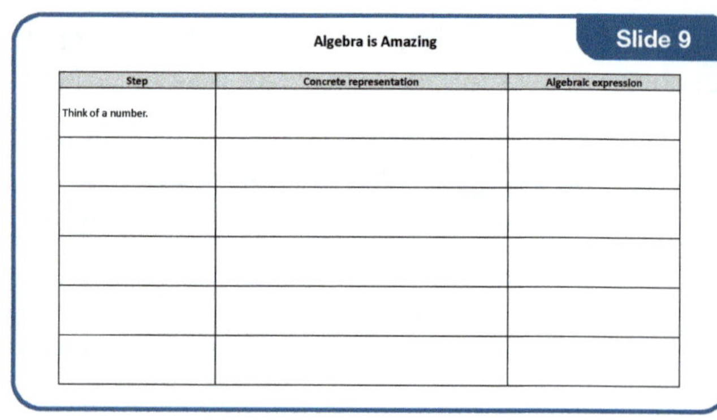

Print slide 9 and distribute to children. You may wish to laminate it and have children write on it with wipe-off pens so that it can be used for different tricks.

You may also wish to combine this task with the simpler 'trick' presented on slide 8 if children need additional support.

Ask children to make a model of each step in the tricks using number rods.

- *Which rod will you use to represent the original 'think of a number'?*

Children may think that they must use the number rod to match the trick's answer. This is not the case. Explain to them they can choose any of the number rods, except the 1-rod.

If children are unsure, model the process with trick A and slide 10.

Slide 10

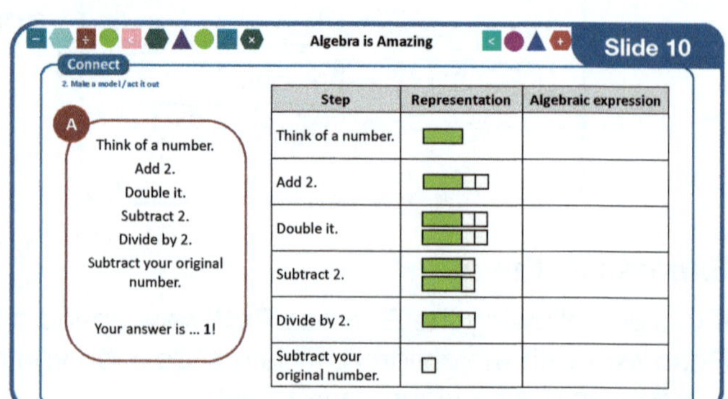

Display slide 10. Talk through each step, asking children to draw the number rods on their copy of the table as you work.

- *When we 'double it' how can we show that with our number rods?*

Check that children focus on replicating the pattern with the rods, not reproducing the number.

⚠ Watch out for children who try to add the number with a single rod, for example, using a 2-rod when adding 2. This makes the pattern harder to identify. Encourage them to use 1-rods as exemplified on the slide.

Once the calculation has been modelled, discuss how to represent each step with algebra.

To help move the children towards the algebraic expressions, you may wish to add a sticky note with an n on to the number rod used to represent the starting number. Together as a class fill in the last column.

Starting number: n

Add 2: $n + 2$

Double it: $2n + 4$

Subtract 2: $2n + 2$

Divide by 2: $n + 1$

Subtract your original number: 1

3. Make a table / spot a pattern

Slide 11

Print slide 11 and distribute to children. This table can be used to support children in migrating to the algebraic solution. Ask them to complete the first table for trick A, with a starting number of 4. Encourage them to record the calculations in extended notation using BIDMAS as appropriate so that they are confident with the calculation. Draw children's attention to the inverse relationships.

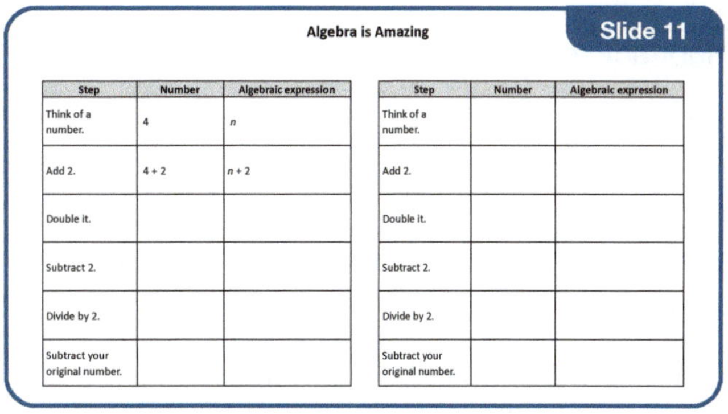

Then ask them to go through the same thought process but replace the original number with n and think about how to record it accurately.

⚠ Children may not be familiar with algebraic notation for multiplication. Explain that:
$2n$ means 2 lots of n, $2 \times n$.
$2(n + 2)$ means 2 lots of $n + 2$; $n + 2 + n + 2 = 2n + 4$

Step	Number	Algebraic expression
Think of a number.	4	n
Add 2.	$4 + 2$	$n + 2$
Double it.	$2(4 + 2)$ $= 2 \times 4 + 4$	$2(n + 2)$ $= 2n + 4$
Subtract 2.	$2 \times 4 + 4 - 2$ $= 8 + 2$	$2n + 4 - 2$ $= 2n + 2$
Divide by 2.	$(8 + 2) \div 2$ $= 4 + 1$	$(2n + 2) \div 2$ $= n + 1$
Subtract your original number.	$4 + 1 - 4$ $= 1$	$n + 1 - n$ $= 1$

Example completed table for trick A

Then ask children to choose a different starting number and use this to complete the second table.

- *What do you notice?*

Slide 12

Print slide 12 and distribute to children for use with tricks B and C.

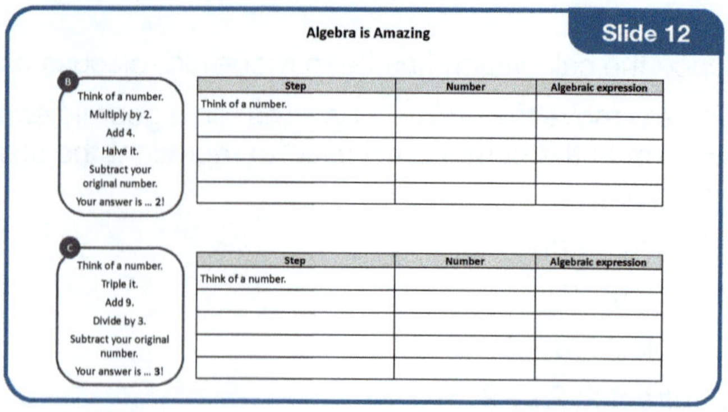

Step	Number	Algebraic expression
Think of a number.	3	n
Multiply by 2.	3×2	$2n$
Add 4.	$3 \times 2 + 4$	$2n + 4$
Halve it.	$(3 \times 2 + 4) \div 2$ $= 3 + 2$	$(2n + 4) \div 2$ $= n + 2$
Subtract your original number.	$3 + 2 - 3$ $= 2$	$n + 2 - n$ $= 2$

Example completed table for trick B

Draw the children's attention to the inverse relationships.

For example, ask:

- *Where is the multiply by 2 'undone'?*

Make sure that children understand that the unknown can be represented by any letter, it does not have to be *n*.

Step	Number	Algebraic expression
Think of a number.	5	s
Triple it.	3×5	$3s$
Add 9.	$3 \times 5 + 9$	$3s + 9$
Divide by 3.	$(3 \times 5 + 9) \div 3$ $= 5 + 3$	$(3s + 9) \div 3$ $= s + 3$
Subtract your original number.	$5 + 3 - 5$ $= 3$	$s + 3 - s$ $= 3$

Example completed table for trick C

Collaborate

Refer to the **Collaborate** task of the **Pupil Resource**.

Ask children to work in pairs or groups to make a set of tricks that have the answers 1 to 10.

- *Think about disguising the language so it is not so easy to spot the inverses. Your aim is to amaze.*
- *Rather than saying multiply by 2 and then in the next step divide by 2, you could say 'double the number' and 'divide by 2'.*

Children could then swap tricks to compare, and work in groups to prepare a presentation as detailed in the **Communicate** task or to investigate other sorts of tricks detailed in the **Create** task.

- *How did you decide which operations to include in your tricks?*
- *Did any of you create the same tricks?*

Communicate

Refer to the **Communicate** task of the **Pupil Resource**.

Slide 13

Display slide 13.

This task provides an opportunity for the children to experience 'performance maths'.

Children should select their best 'trick' and perform it to others, such as another class, governors or senior leaders.

Performances could be live or recorded. Some children may not be comfortable performing; they may prefer to support others.

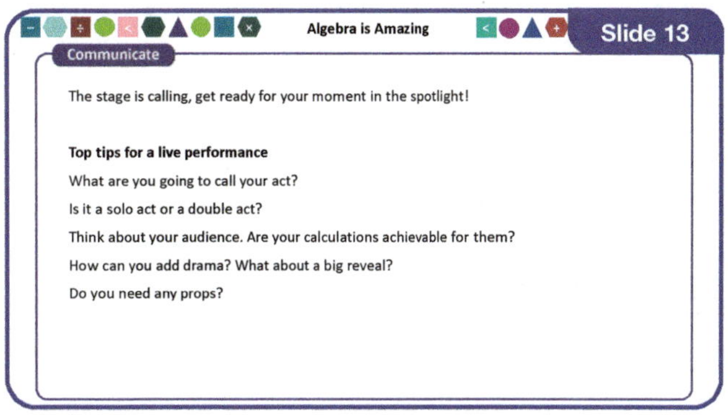

- *What are you going to call your act? Is it a solo act or a double act?*
- *Think about your audience. Are your calculations achievable for them?*
- *How can you add drama (for example, hide the number somewhere for a big reveal)?*
- *Do you need any props?*

Create

Refer to the **Create** task of the **Pupil Resource**.

Slide 14

Display slide 14 which contains an example of other tricks that are possible.

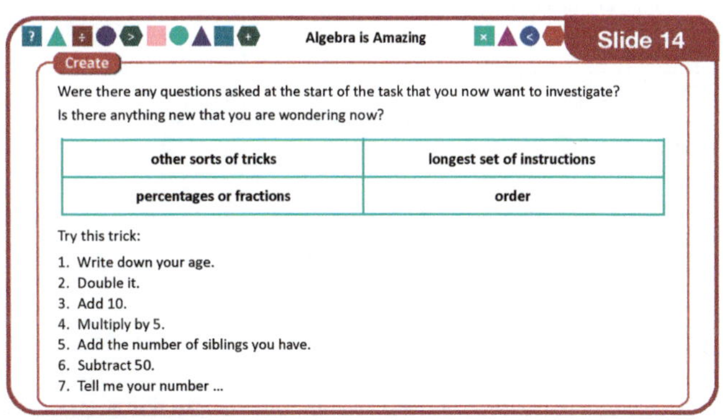

- *Were there any questions asked at the start of the task that you now want to investigate?*
- *Is there anything new that you are wondering now?*

Children can use the word bank to prompt further questions of their own.

other sorts of tricks	**longest set of instructions**
percentages or fractions	**order**

- I wonder if there is a set of instructions which can be changed to make any answer possible.
- I wonder if I will still get the same answer if I change the order in which I follow the instructions.
- I wonder if I can make the longest set of instructions that still works.
- I wonder whether I can include other calculations, such as percentages.
- I wonder if I can solve this trick.

Consider

Refer to the **Consider** task of the **Pupil Resource.**

Slide 15

Display slide 15.

Encourage children to think about the strategies that they have used and those that their peers have described in their presentations. The prompts below may help get children started.

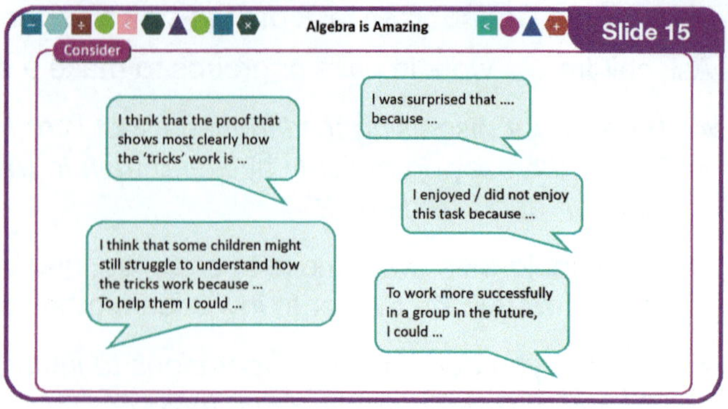

Cognitive learning domain

- I think that the proof that shows mostly clearly how the 'tricks' work is …
- The new strategies that I learned during this investigation were ... This changed my thinking because ...
- I think that some children might still struggle to understand how the tricks work because … To help them I could …

Affective domain

- I was surprised that … because …
- I was influenced by … and as a result …
- I enjoyed / did not enjoy this task because …

Metacognitive domain

- I found … helpful.
- I found … unhelpful.
- To work more successfully as a member of a group in the future, I could …

Answers

Trick A

Step	Algebraic expression
Think of a number.	n
Add 2.	$n + 2$
Double it.	$2(n + 2)$
Subtract 2.	$2n + 2$
Divide by 2.	$n + 1$
Subtract your original number.	1

Trick B

Step	Algebraic expression
Think of a number.	n
Multiply by 2.	$2n$
Add 4.	$2n + 4$
Halve it.	$n + 2$
Subtract your original number.	2

Trick C

Step	Algebraic expression
Think of a number.	n
Triple it.	$3n$
Add 9.	$3n + 9$
Divide by 3.	$n + 3$
Subtract your original number.	3

A trick to make people think of the number 4

For example:

Step	Algebraic expression
Think of a number.	n
Triple it.	$3n$
Add 12.	$3n + 12$
Divide by 3.	$n + 4$
Subtract your original number.	4

Ancient Egyptian Fractions

Year 5
Year 6

Introduction

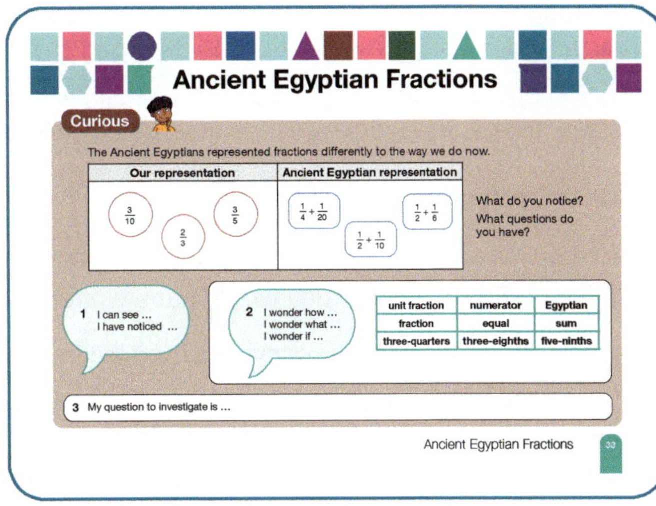

Ancient Egyptian Fractions allows children to apply their knowledge of equivalent fractions and calculating with fractions to an interesting historical context. To get the most out of the task, children should be confident working with unit and non-unit fractions.

Children make models and draw diagrams to explore and explain the abstract calculations in a deep conceptual way. This reinforces that building mathematical models and representing ideas pictorially is a beneficial strategy at all stages of mathematical learning.

This task promotes greater-depth learning in a range of ways, supporting children to identify and articulate underlying mathematical structures. For example, in the **Communicate** stage, greater depth is promoted by asking children to develop clear, accurate and well-organised communication, with ideas flowing from one to the next, and illustrated using diagrams or models as needed.

This **Teacher Guide** is supported by photocopiable **Pupil Resources** and **Teacher's Presentation** PowerPoint slides for whole-class or group display.

National Curriculum

Ancient Egyptian Fractions covers these Year 5 and Year 6 National Curriculum objectives.

Year 5 Fractions

- Compare and order fractions whose denominators are all multiples of the same number.
- Identify, name and write equivalent fractions of a given fraction, represented visually, including tenths and hundredths.

Year 6 Fractions

- Use common factors to simplify fractions; use common multiples to express fractions in the same denomination.
- Compare and order fractions, including fractions >1.

The 6C cycle

The grid on the right shows the relative weighting given to each of the 6 Cs in **Ancient Egyptian Fractions**.

Curious	★★
Connect	★★
Collaborate	★★
Communicate	★★★
Create	★★
Consider	★

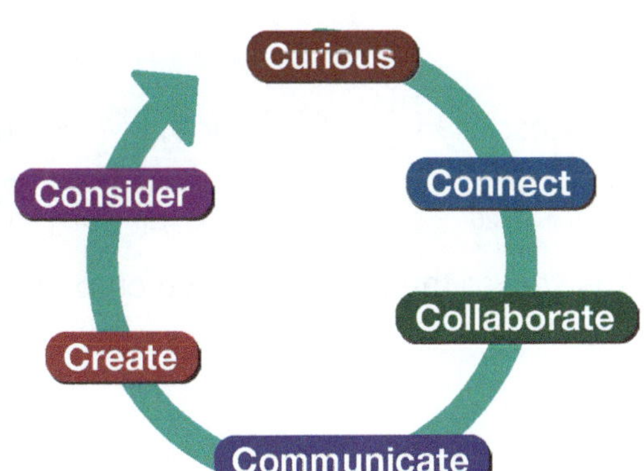

Problem-solving heuristics

The grid shows which problem-solving strategies children are likely to use when investigating the **Connect** tasks in **Ancient Egyptian Fractions**. These strategies are further exemplified in the **Connect** stage below (pages 101–102).

Draw a diagram	✓
Make a table / spot a pattern	
Make a model / act it out	✓
Try then improve	
Create a systematic list	
Solve a simpler but related problem	
Work backwards	✓

Key mathematical ideas

Ancient Egyptians wrote fractions as a sum of unit fractions. All the denominators in the sum had to be different. For example, they would have written $\frac{2}{3}$ as $\frac{1}{2} + \frac{1}{6}$.

This task offers two ways to convert a modern fraction, for example, $\frac{3}{5}$, to Ancient Egyptian fractions.

Firstly, within a practical, sharing context. For example, share 3 tray bake cakes between 5 people. The aim is to use as few fractions as possible, so we start with the largest possible unit fraction. Can we give each person $\frac{1}{2}$ of a tray bake? (Yes.) How much is left over? ($\frac{1}{2}$ of 1 tray bake.)

Share that final half between the 5 people. Each person gets $\frac{1}{5}$ of the left over $\frac{1}{2}$, which is $\frac{1}{10}$ of a cake.

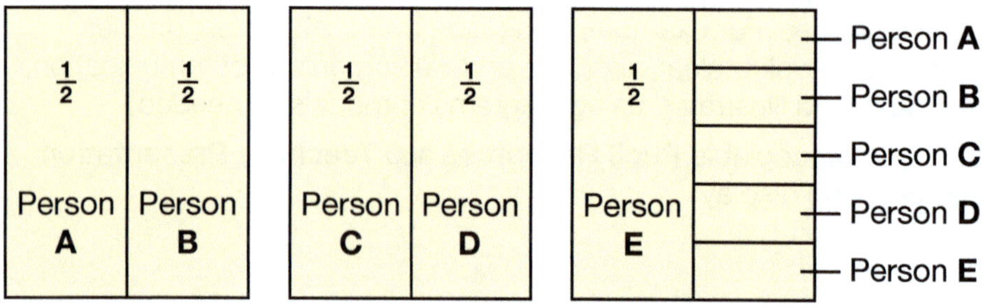

Each person gets $\frac{1}{2}$ and $\frac{1}{10}$.

$\frac{3}{5} = \frac{1}{2} + \frac{1}{10}$

Secondly, with a method based on the 'greedy algorithm'. This strategy, developed by Fibonacci, involves subtracting the largest possible unit fraction each time. The first step is to establish whether the fraction you are starting with is greater or less than the largest unit fraction, $\frac{1}{2}$.

For example, is $\frac{3}{5}$ greater or less than $\frac{1}{2}$? (Greater.) So can $\frac{1}{2}$ be subtracted? (Yes.)

$\frac{3}{5} - \frac{1}{2} = \frac{1}{10}$ $\quad$ $\frac{6}{10} - \frac{1}{2} = \frac{6}{10} - \frac{5}{10} = \frac{1}{10}$ $\quad$ $\frac{3}{5}$ as an Ancient Egyptian fraction is $\frac{1}{2} + \frac{1}{10}$.

If the starting fraction is less than $\frac{1}{2}$, move to the next largest unit fraction, $\frac{1}{3}$. For example:

$\frac{2}{5} - \frac{1}{3} = \frac{1}{15}$ $\quad$ $\frac{6}{15} - \frac{1}{3} = \frac{6}{15} - \frac{5}{15} = \frac{1}{15}$ $\quad$ $\frac{2}{5}$ as an Ancient Egyptian fraction is $\frac{1}{3} + \frac{1}{15}$

Sometimes you will need to continue subtracting until all numerators are 1. For example:

$\frac{4}{5} - \frac{1}{2} = \frac{3}{10}$ $\quad$ $\frac{8}{10} - \frac{1}{2} = \frac{8}{10} - \frac{5}{10} = \frac{3}{10}$

$\frac{3}{10}$ is less than $\frac{1}{3}$ but greater than $\frac{1}{4}$ so subtract $\frac{1}{4}$ next.

$\frac{3}{10} - \frac{1}{4} = \frac{1}{20}$ $\quad$ $\frac{6}{20} - \frac{1}{4} = \frac{6}{20} - \frac{5}{20} = \frac{1}{20}$ $\quad$ Therefore, $\frac{4}{5} = \frac{1}{2} + \frac{1}{4} + \frac{1}{20}$.

Answers to the tasks are given on pages 104–105.

Teaching guidance

If you are using the **Express Pathway**, please turn to pages 100–104.

Teaching guidance for the **Full 6C Pathway** is given below.

The **Teacher's Presentation** PowerPoint file has slides specifically mentioned in this teaching guidance. Some of the slides include notes (within the PowerPoint file) providing 'stumble support' suggestions, that may help you to support children in their problem-solving thinking and enable access for children who are less confident.

Awe and wonder

Slide 1

Display slide 1.

Ask children to share what they are noticing or wondering.

Our representations:

- I can see fractions.
- I can see that they all have numerators greater than one.
- I have noticed that two of the fractions have the same numerator.

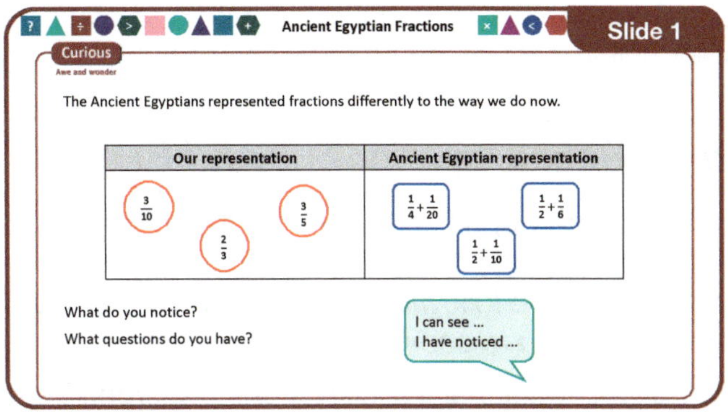

Ancient Egyptian fractions:

- I can see that they have the same numerator, which is 1.
- I can see that they are addition calculations without answers.
- I have noticed that all the fractions in each sum have different denominators.
- I have noticed that the fractions in each calculation get smaller.

⚠ A common error with adding fractions is to add both the numerator and the denominator. To check this, write on the board: $\frac{1}{7}+\frac{1}{7}=\frac{2}{7}$ and $\frac{1}{7}+\frac{1}{7}=\frac{2}{14}$. Ask children to tell you which one is correct and why.

Develop and deepen

Slide 2

Display slide 2. Explain the rules for Ancient Egyptian fractions.

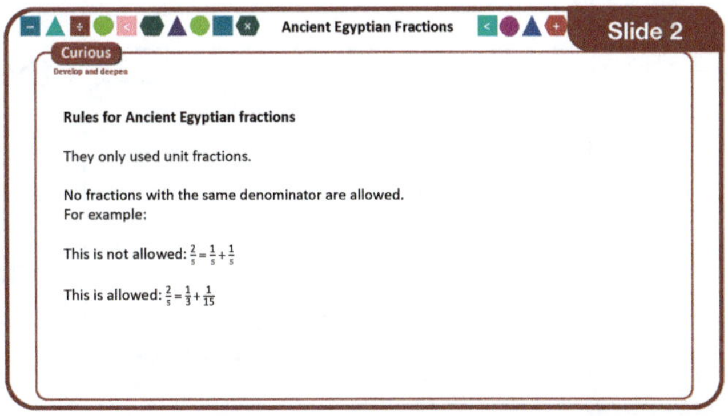

Slide 3

Display slide 3.

Encourage children to use the words in the word bank to develop and deepen their thoughts and questions.

unit fraction	numerator	Egyptian
fraction	equal	sum
three-quarters	three-eighths	five-ninths

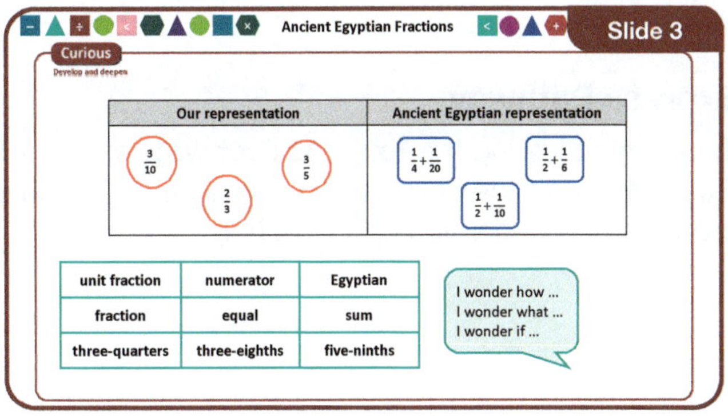

Possible lines of enquiry:

- I wonder how this works.

Possible lines of enquiry linked to the word bank:

- I wonder if I can match the Ancient Egyptian representation to the way we write fractions.
- I wonder if there are any fractions that cannot be written as the sum of unit fractions.

Possible lines of enquiry that children may raise now, but should be held back until the **Create** task:

- I wonder if I could write a set of matching fractions, and challenge someone to match them up.
- I wonder what fractions I can make by adding consecutive unit fractions.

Consider whether / how to collate and display all the suggested questions, and whether / how to steer children towards selecting an appropriately challenging question.

Pupil Resources for Full and Express Pathways

Distribute the **Pupil Resources** for **Ancient Egyptian Fractions** to the children to help support them in their exploration of the different tasks. Use either of the **Curious Pupil Resources** depending on whether you are following the **Full Pathway** or the **Express Pathway**. The **Pupil Resources** for the remaining 5Cs are common to both pathways.

To provide additional flexibility in the classroom, each of the **Pupil Resources** is also provided in the **Teacher's Presentation** PowerPoint file (see slides 4–7), so you can display these at any time if you wish.

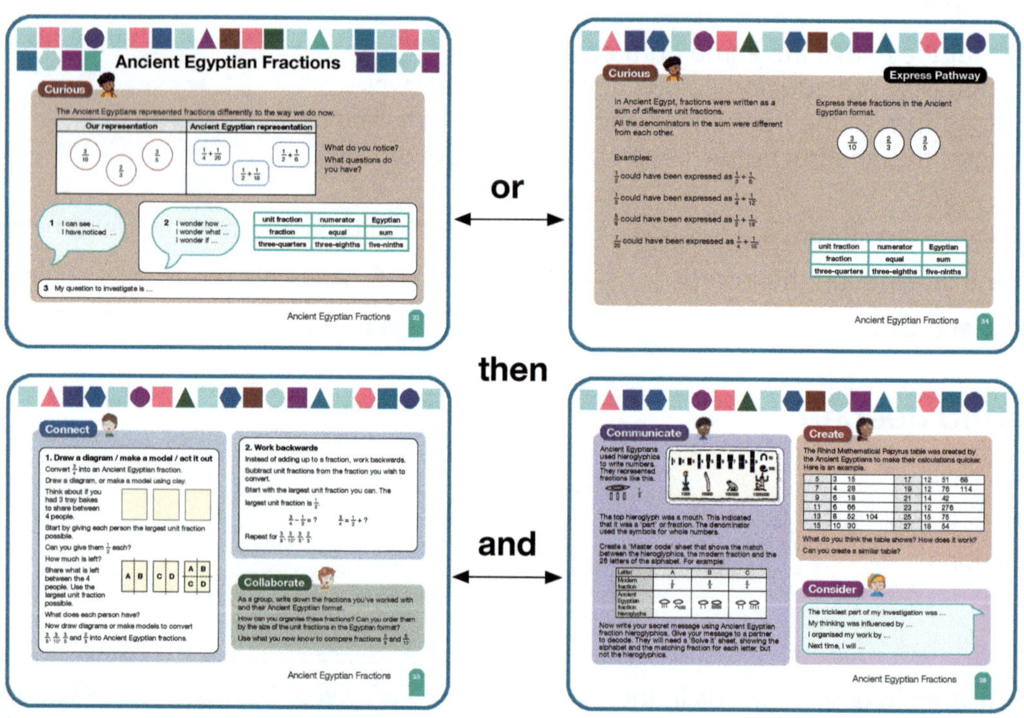

Express Pathway

If you are following the **Express Pathway**, spend a few minutes discussing the **Curious Pupil Resource** with the children. (This is slide 5, so you could display it if you wish.) Use some of the prompts on the previous page to help.

Two tasks involving different strategies are suggested for children to make connections.

1. Draw a diagram / make a model / act it out

Additional resources required: paper, pencils, modelling clay, clay tools

Slide 8

Display slide 8.

- *Consider what three-quarters means. Think about if you had 3 tray bakes to share between 4 people.*
- *Start by giving each person the largest unit fraction possible. What is the largest unit fraction?* ($\frac{1}{2}$)

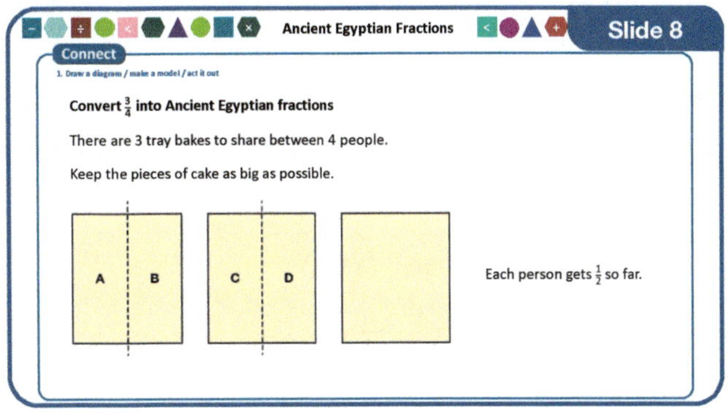

Children can model this by drawing three rectangles and dividing the first two in half, or by making three rectangles out of clay and cutting the first two in half.

- *Can I give $\frac{1}{2}$ a tray bake to each person?* (Yes.)
- *How much is left?* (One whole tray bake.)
- *What's the next largest unit fraction?* ($\frac{1}{3}$)
- *If we share the leftover tray bake into thirds, can we give each person 1 piece?* (No.)

Slide 9

Display slide 9.

- *What's the next largest unit fraction?* ($\frac{1}{4}$)
- *If we share it into quarters, will we have enough to give each person 1 piece?* (Yes.)
- *How much cake is left over?* (None.)
- *How much cake did each person get?* ($\frac{1}{2} + \frac{1}{4} = \frac{3}{4}$)

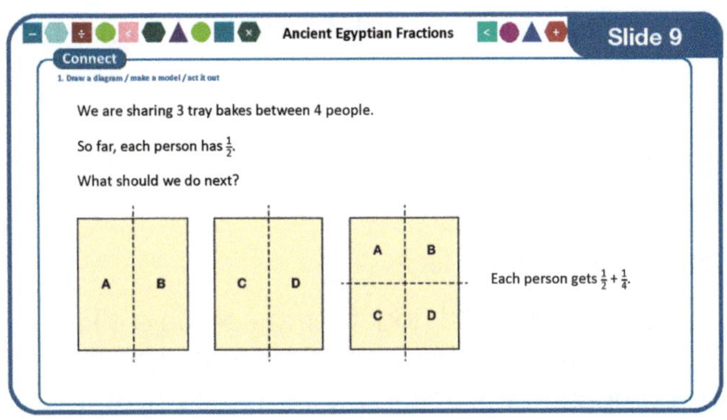

Slide 10

Display slide 10.

Ask children to convert $\frac{3}{8}$ using modelling clay or diagrams.

- If I have 3 tray bakes to share between 8 people, I can't give everyone half – there's not enough.
- I can give everyone one-third and then there is one-third left over. I can split this final third into eighths.
- Each person gets $\frac{1}{3} + \frac{1}{24} = \frac{3}{8}$.

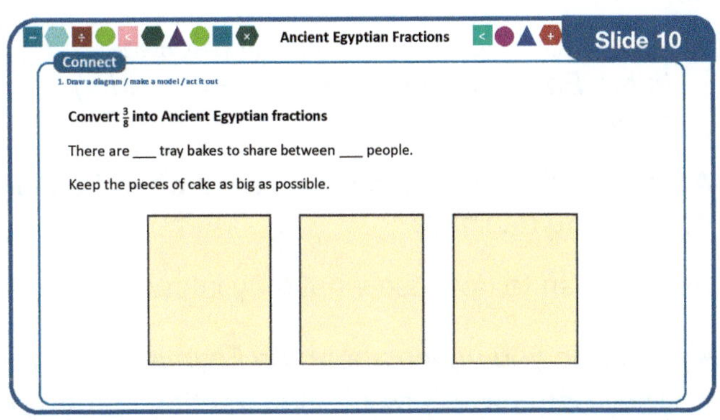

Agree that $\frac{3}{8}$ as an Ancient Egyptian fraction is $\frac{1}{3} + \frac{1}{24}$.

Ask children to also draw diagrams to check the conversions for $\frac{3}{10}$, $\frac{2}{3}$ and $\frac{3}{5}$ from the **Curious** task.

Connect

2. Work backwards

Explain that, instead of adding the unit fractions together, children can try subtracting unit fractions instead. Tell children that this method is called the 'greedy algorithm', which was discovered by Fibonacci.

Working systematically, children should write down the largest possible unit fraction that is smaller than the fraction they are looking to convert.

Slide 11

Display slide 11. Talk through the two examples.

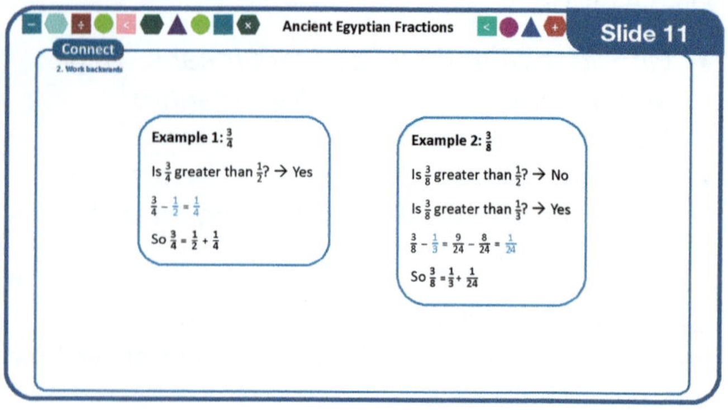

Example 1: $\frac{3}{4}$

- *Is it larger than half?* (Yes.)

$\frac{3}{4} - \frac{1}{2} = \frac{1}{4}$ so: $\frac{3}{4} = \frac{1}{2} + \frac{1}{4}$

Example 2: $\frac{3}{8}$

- *Is it larger than half?* (No.)
- *Is it larger than third?* (Yes.)

$\frac{3}{8} - \frac{1}{3} = \frac{9}{24} - \frac{8}{24} = \frac{1}{24}$ so: $\frac{3}{8} = \frac{1}{3} + \frac{1}{24}$

Encourage children to use benchmarking principles to help them decide the largest unit fraction to use.

Once children have explored working backwards for $\frac{3}{4}$ and $\frac{3}{8}$, ask them to use this method to check the matches for $\frac{3}{10}$, $\frac{2}{3}$ and $\frac{3}{5}$ from the **Curious** task.

Collaborate

Refer to the **Collaborate** task of the **Pupil Resource**.

Slide 12

Display slide 12.

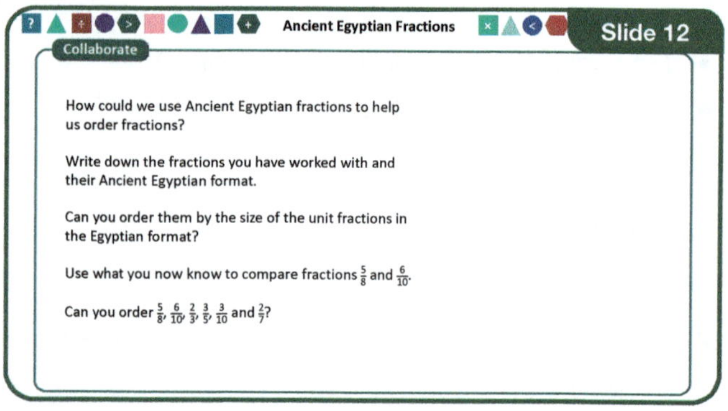

Ask children to work together to compare all the fractions they have managed to find so far, by comparing the unit fractions in the Ancient Egyptian format.

- *When could looking at the unit fractions in the Egyptian format help when ordering fractions?*

Agree that this method can be useful when fractions are very close together and they do not easily convert into decimals.

Ask children to now use what they know to compare the fractions $\frac{5}{8}$ and $\frac{6}{10}$.

- *What do you notice about the Egyptian versions of these two fractions?* ($\frac{5}{8} = \frac{1}{2} + \frac{1}{8}$; $\frac{6}{10} = \frac{1}{2} + \frac{1}{10}$)
- They both have $\frac{1}{2}$ as the first fraction in the sum of unit fractions, so we have to compare the second unit fraction in the sum (the second addend).
- *Which is greater, $\frac{1}{8}$ or $\frac{1}{10}$?*
- $\frac{1}{8}$ is greater than $\frac{1}{10}$. Therefore $\frac{5}{8}$ is greater than $\frac{6}{10}$.

Refer to the **Communicate** task of the **Pupil Resource**.

Slide 13

Children could create their own codes based on Ancient Egyptian fractions and write some secret messages.

Display slide 13. Explain how the Ancient Egyptians represented their numbers and fractions.

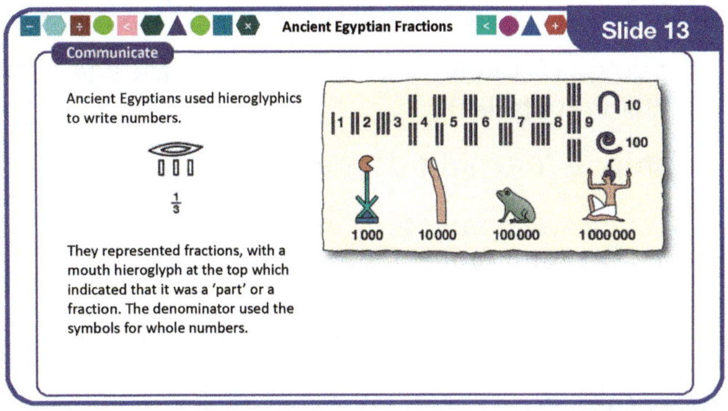

Slide 14

Print slide 14 and distribute to children. They should use this to create their master code sheet. They choose a fraction for each letter, work out the Ancient Egyptian fraction equivalent and draw the hieroglyphics.

Ask children to then write their secret message using the Ancient Egyptian fraction hieroglyphics.

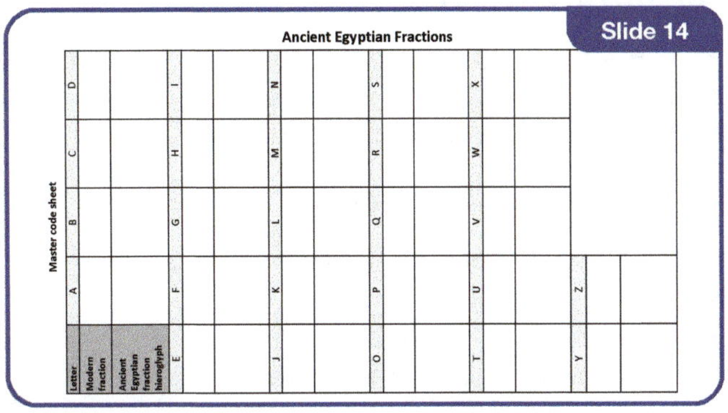

They should then pass their message to a partner, along with a 'Solve it' sheet that shows the alphabet and the matching fraction, but not the hieroglyphics.

Letter	A	B	C
Modern fraction	$\frac{2}{5}$	$\frac{5}{8}$	$\frac{3}{4}$
Ancient Egyptian fraction hieroglyphs			

Example code sheet

Refer to the **Create** task of the **Pupil Resource**.

- *Were there any questions asked at the start of the task that you now want to investigate?*
- *Is there anything new that you are wondering now?*
- I wonder if I could write a set of matching fractions, and challenge someone to match them up.
- I wonder what fractions I can make by adding consecutive unit fractions.
- I wonder if there is a pattern when the numerator is 2.

Slide 15

Display slide 15.

Explain that the Rhind Mathematical Papyrus table was created by the Ancient Egyptians to help them with their calculations.

- *What do you think the table shows? How does it work?*

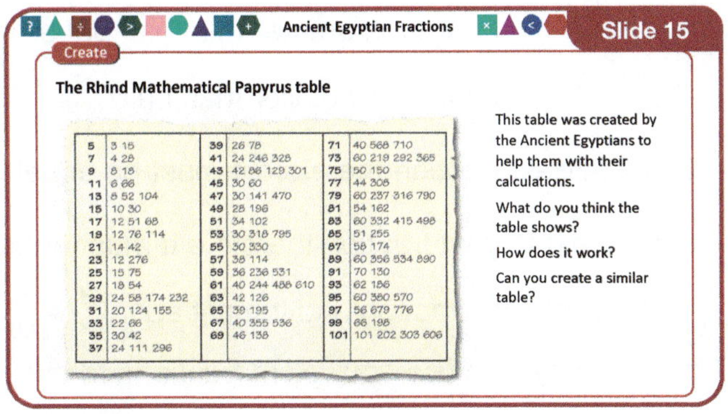

After discussion, explain that this table represents how to express fractions of the form $\frac{2}{n}$ where n is an odd number from 5 to 101.

For example, the first entry shows that $\frac{2}{5} = \frac{1}{3} + \frac{1}{15}$.

Ask children to create a similar (smaller) table, but for fractions of the form $\frac{3}{n}$.

Consider

Refer to the **Consider** task of the **Pupil Resource**.

Encourage children to think about the strategies that they have used. The prompts below may help get children started.

Cognitive learning domain

- The trickiest part of my investigation was …
- I think that I have / have not identified similar things to others in my group. I think that this is because …
- Now that I have completed this task, I feel more / less confident. This is because …

Affective domain

- The fun bit of my learning was … It was fun because …
- My thinking was influenced by …
- I was surprised about … because …

Metacognitive domain

- I used … to help me. I was helpful because …
- I found … unhelpful.
- I organised my work by …
- I worked well with others by …
- Next time, I will …

Answers

Curious

$\frac{3}{10}$ in Ancient Egyptian fraction form is $\frac{1}{4}+\frac{1}{20}$

$\frac{2}{3}$ in Ancient Egyptian fraction form is $\frac{1}{2}+\frac{1}{6}$

$\frac{3}{5}$ in Ancient Egyptian fraction form is $\frac{1}{2}+\frac{1}{10}$

The proof that children produce may vary.

During the task children may talk about the approximate magnitude of the fractions:

$\frac{1}{2}+\frac{1}{6}$	answer is greater than half
$\frac{1}{2}+\frac{1}{10}$	answer is greater than half but smaller than $\frac{1}{2}+\frac{1}{6}$
$\frac{1}{4}+\frac{1}{20}$	answer is smaller than half

$\frac{2}{3}$ is the largest fraction therefore it must be equal to $\frac{1}{2}+\frac{1}{6}$.

$\frac{3}{5}$ is the next largest fraction therefore it must equal $\frac{1}{2}+\frac{1}{10}$.

$\frac{3}{10}$ is the smallest fraction and it must equal $\frac{1}{4}+\frac{1}{20}$.

Children may also use equivalent fractions to show that their answers are correct.

$\frac{1}{2}+\frac{1}{6}=\frac{3}{6}+\frac{1}{6}=\frac{4}{6}=\frac{2}{3}$

$\frac{1}{2}+\frac{1}{10}=\frac{5}{10}+\frac{1}{10}=\frac{3}{5}$

$\frac{1}{4}+\frac{1}{20}=\frac{5}{20}+\frac{1}{20}=\frac{6}{20}=\frac{3}{10}$

Example table based on the Rhind Mathematical Papyrus table

4	2	4
5	2	10
6	2	
7	–	
8	3	24
9	3	
10	4	20

Workings:

$\frac{3}{4} = \frac{1}{2} + \frac{1}{4}$

$\frac{3}{4}$ is greater than $\frac{1}{2}$, so subtract this first.

$\frac{3}{4} - \frac{1}{2} = \frac{3}{4} - \frac{2}{4} = \frac{1}{4}$ so: $\frac{3}{4} = \frac{1}{2} + \frac{1}{4}$

$\frac{3}{5} = \frac{1}{2} + \frac{1}{10}$

$\frac{3}{5}$ is greater than $\frac{1}{2}$, so subtract this first.

$\frac{3}{5} - \frac{1}{2} = \frac{6}{10} - \frac{5}{10} = \frac{1}{10}$ so: $\frac{3}{5} = \frac{1}{2} + \frac{1}{10}$

$\frac{3}{6} = \frac{1}{2}$

Equivalent unit fraction

$\frac{3}{7} = ?$

$\frac{3}{7}$ is smaller than $\frac{1}{2}$ but greater than $\frac{1}{3}$, so subtract this first.

$\frac{3}{7} - \frac{1}{3} = \frac{9}{21} - \frac{7}{21} = \frac{2}{21}$

As this does not fit the pattern, children are likely to record that they cannot find a pair of unit fractions to equal $\frac{3}{7}$.

$\frac{3}{8} = \frac{1}{3} + \frac{1}{24}$

$\frac{3}{8}$ is smaller than $\frac{1}{2}$ but greater than $\frac{1}{3}$, so subtract this first.

$\frac{3}{8} - \frac{1}{3} = \frac{9}{24} - \frac{8}{24} = \frac{1}{24}$ so: $\frac{3}{8} = \frac{1}{3} + \frac{1}{24}$

$\frac{3}{9} = \frac{1}{3}$

Equivalent unit fraction

$\frac{3}{10} = \frac{1}{4} + \frac{1}{20}$

$\frac{3}{10}$ is smaller than $\frac{1}{3}$ but greater than $\frac{1}{4}$, so subtract this first.

$\frac{3}{10} - \frac{1}{4} = \frac{6}{20} - \frac{5}{20} = \frac{1}{20}$ so: $\frac{3}{10} = \frac{1}{4} + \frac{1}{20}$

Puzzling Percentages

Year 6

Introduction

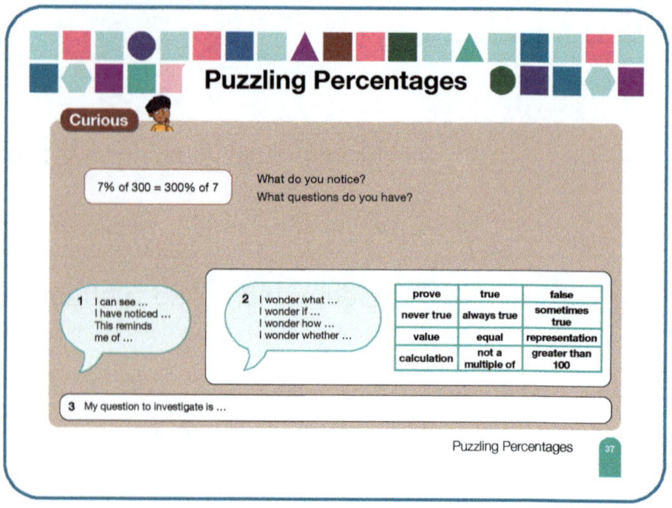

Puzzling Percentages provides children with an opportunity to deepen their understanding of the connections between percentages and multiplicative structures. Children explore the commutativity of $a\%$ of $b = b\%$ of a, using calculation skills and their prior experience to show that the equation is true.

Through the **Connect** stage, children will consider how to represent the problems and explain their choices confidently. Connections are also made to fractions with children exploring testing whether similar commutativity rules apply.

This task promotes greater-depth learning in a range of ways, supporting children to identify and articulate underlying mathematical structures. For example, in the **Curious** stage, greater depth is promoted by asking children to make rich mathematical contributions to discussions and suggest a variety of appropriate mathematical questions.

This **Teacher Guide** is supported by photocopiable **Pupil Resources** and **Teacher's Presentation** PowerPoint slides for whole-class or group display.

National Curriculum

Puzzling Percentages covers these Year 6 National Curriculum objectives.

Fractions (including decimals and percentages)

- Recall and use equivalences between simple fractions, decimals and percentages, including in different contexts.

Ratio and proportion

- Solve problems involving the calculation of percentages [for example, of measures, and such as 15% of 360] and the use of percentages for comparison.

The 6C cycle

The grid on the right shows the relative weighting given to each of the 6 Cs in **Puzzling Percentages**.

Curious	★★★
Connect	★★★
Collaborate	★★★
Communicate	★★
Create	★★
Consider	★

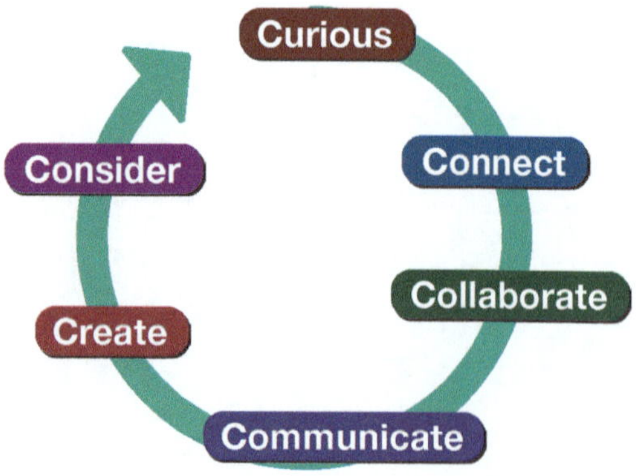

Problem-solving heuristics

The grid shows which problem-solving strategies children are likely to use when investigating the **Connect** tasks in **Puzzling Percentages.** These strategies are further exemplified in the **Connect** stage below (pages 110–113).

Draw a diagram	✓
Make a table / spot a pattern	
Make a model / act it out	✓
Try then improve	
Create a systematic list	
Solve a simpler but related problem	✓
Work backwards	

Key mathematical ideas

Puzzling Percentages allows children to deepen their understanding of the commutativity of multiplication, for example, that 3×4 ($a \times b$) $= 4 \times 3$ ($b \times a$), by asking them to investigate the statement a% of $b = b$% of a.

In order to get the most out of the task, children should be confident in the fact that 'per cent' means the number of parts per hundred, and that 100% represents the whole quantity. Children will also benefit from an understanding of the relationship between percentages, fractions and decimals, for example, that 50% is equal to half or 0·5 / $\frac{50}{100}$ of the whole quantity and 25% is equal to a quarter or 0·25 / $\frac{25}{100}$ of the whole quantity.

This task asks children to find 7% of 300 and 300% of 7 and compare the amounts. If children have not worked with percentages greater than 100% before, you may like to discuss the fact that a percentage larger than 100% can be treated in the same way as one smaller than 100%, and like an improper fraction will be more than one whole.

There are a number of strategies for finding percentages suitable for this task, such as finding 1% of 300 and multiplying by 7 or by applying an understanding of scaling:

- 7% of 100 equals 7, and 7% of 300 is equal to 21 (multiplying both by 3);
- using knowledge that 300% is 3 times as much as 100% to work out that 300% of 7 is 21, for example, '100% of 7 is 7, 200% of 7 is 14 and 300% of 7 is 21.'

Answers are given in situ throughout the tasks and are also summarised on page 115.

Teaching guidance

If you are using the **Express Pathway**, please turn to pages 110–115.

Teaching guidance for the **Full 6C Pathway** is given below.

The **Teacher's Presentation** PowerPoint file has slides specifically mentioned in this teaching guidance. Some of the slides include notes (within the PowerPoint file) providing 'stumble support' suggestions, that may help you to support children in their problem-solving thinking and enable access for children who are less confident.

Awe and wonder

Reveal slides 1–2 one at a time, allowing time for discussion and observations as prompted below.

Slide 1

Display slide 1.

- I can see that 7 and 300 are used twice.
- I can see that the percentage symbol appears twice.
- I have noticed that the equals sign is in the middle of the equation / number sentence.
- I have noticed that 300% is greater than 100%. This might mean 3 wholes.
- This reminds me of the fact that addition can be done any order, for example $10 + 3 = 3 + 10$.

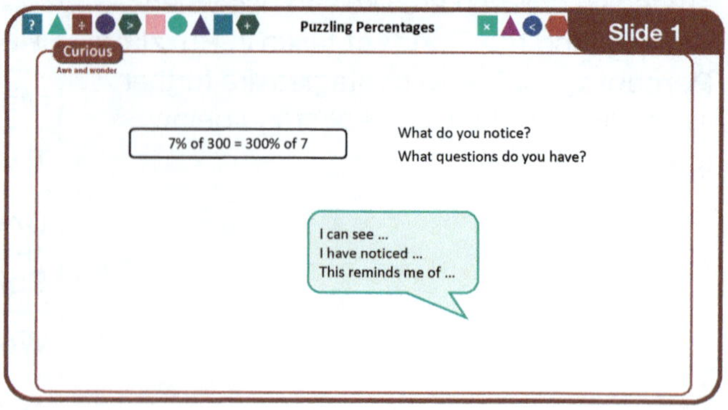

Slide 2

Dependent on children's confidence, you may choose to replace 'of' with the multiplication symbol and display the prompt $7\% \times 300 = 300\% \times 7$ on Slide 2.

- *What's the same? What's different?*
- I have noticed that the multiplication symbol is used instead of the word 'of'.

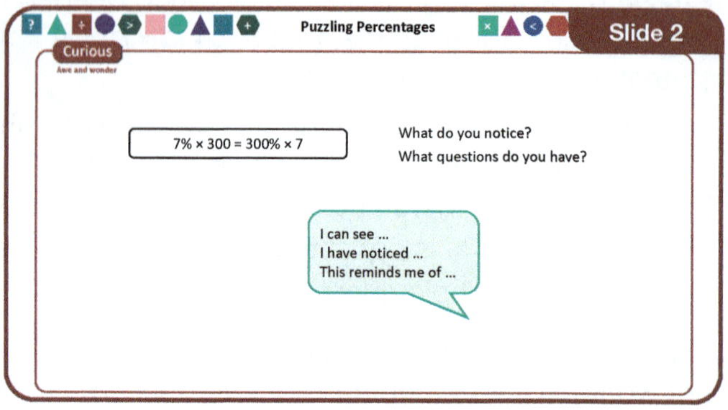

⚠ Children may not have encountered the following before:

- percentages over 100%;
- the multiplication symbol being used to represent 'of'.

Children could be introduced to the idea of percentages greater than 100% through concrete resources that allow them to construct 100% (or one whole) and then continue to build the pattern. For example, show a base-10 hundreds-block.

- *If this represents 100%, what would 1 hundreds-block and 1 tens-rod be worth?* (110%)

Continue building until you reach 200% (2 hundreds-blocks).
Then move to numerical calculations and bar models.

- *If 100% is worth 50, what is 10%?* (5)
- *What is 110%?* (55)

Develop and deepen

Encourage children to use the words in the word bank to develop and deepen their thoughts and questions.

prove	**true**	**false**
never true	**always true**	**sometimes true**
value	**equal**	**representation**
calculation	**not a multiple of**	**greater than 100**

Possible lines of enquiry:

- I wonder what the value of each calculation is.
- I wonder if the multiplication symbol means the same as 'of'. [Depending if slide 2 has been displayed.]
- I wonder whether this number sentence is true or false.

Possible lines of enquiry linked to the word bank:

- I wonder how I can prove that this works.
- I wonder what different representations I can use to prove this is true or false.
- I wonder if this is always true for all percentages or just sometimes.
- I wonder if it works when neither of the numbers are a multiple of 10.

Possible lines of enquiry that children may raise now, but should be held back until the **Create** task:

- I wonder if this works for fractions and decimals.

Consider whether / how to collate and display all the suggested questions, and whether / how to steer children towards selecting an appropriately challenging question.

Pupil Resources for Full and Express Pathways

Distribute the **Pupil Resources** for **Puzzling Percentages** to the children to help support them in their exploration of the different tasks. Use either of the **Curious Pupil Resources** depending on whether you are following the **Full Pathway** or the **Express Pathway**. The **Pupil Resources** for the remaining 5Cs are common to both pathways.

To provide additional flexibility in the classroom, each of the **Pupil Resources** is also provided in the **Teacher's Presentation** PowerPoint file (see slides 3–6), so you can display these at any time if you wish.

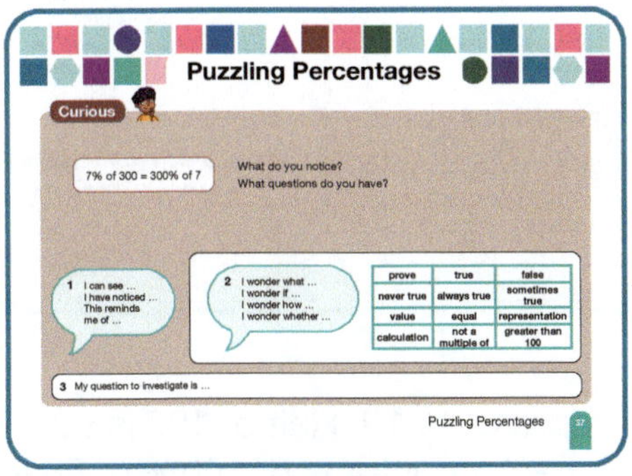

or

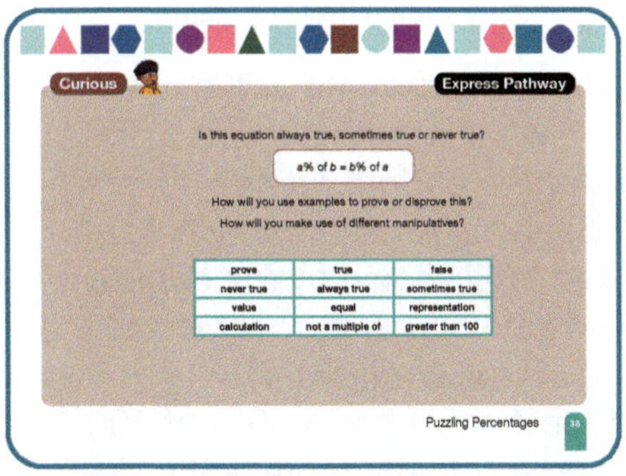

then

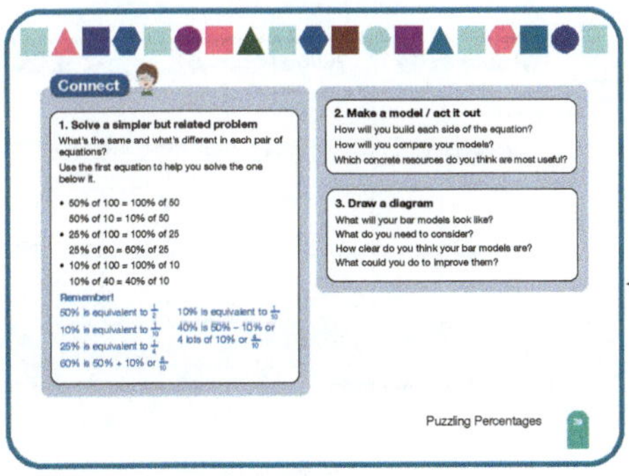

and

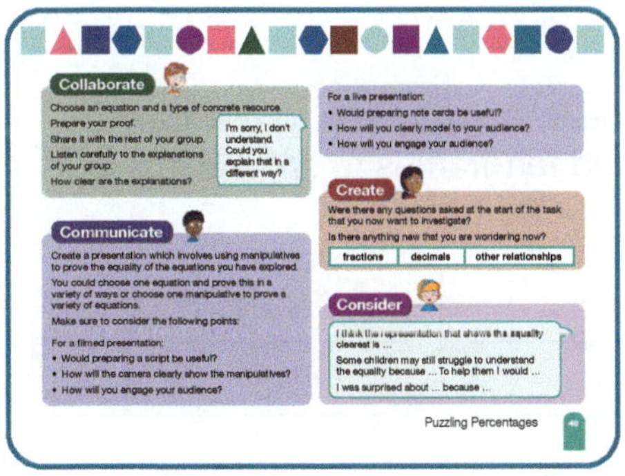

Express Pathway

If you are following the **Express Pathway**, spend a few minutes discussing the **Curious Pupil Resource** with the children. (This is slide 4, so you could display it if you wish.) Use some of the prompts on the previous page to help.

If using the **Express Pathway**, when asked what they would like to explore next, prompt children to consider fractions if they do not come up with this line of enquiry themselves.

Three tasks involving different strategies are suggested for children to make connections.

1. Solve a simpler but related problem

Additional resources required: ten-frames, counters, whiteboard pens (optional)

Slide 7

Display slide 7. Ask children to say whether the equations are true. Encourage them to use the information in blue to help them, and to use what they know about the first equation to help them with the one below it.

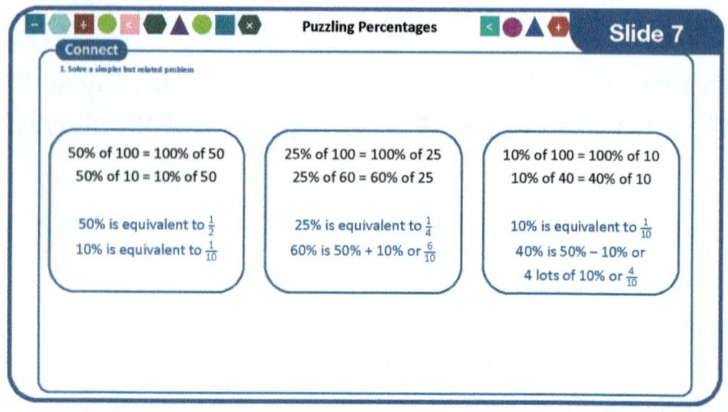

Slide 8

Display slide 8 to model proving the equation 50% of 100 = 100% of 50. Make ten-frames and counters available so that children may build their own version. For additional support they may wish to annotate the counters using whiteboard pens to prove the equality.

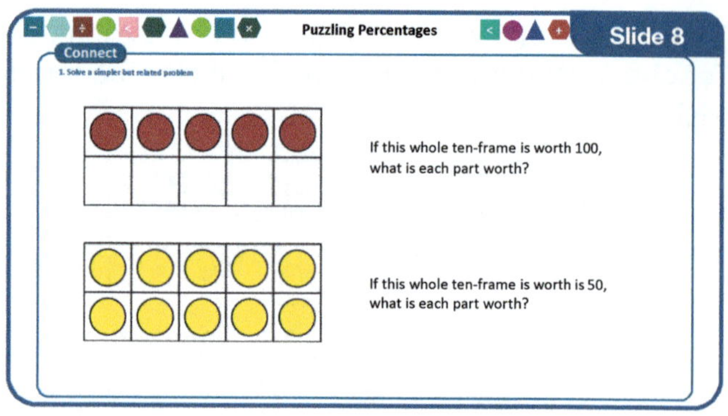

- *Look at the top ten-frame. If the whole is worth 100, what is each part worth?* (10)
- *What about the bottom frame? If the whole is worth 50, what is each part worth?* (5)
- *What is the total shown by the counters in each ten-frame? What percentage of the ten-frame do the counters fill?*
- The first ten-frame shows 50% of 100. Each counter represents 10. Half of 100 is 50.
- The second ten-frame represents 50 and is 100% full. Each counter is 5. 100% of 50 is 50.

Slide 9

Display slide 9. Ask children to use their counters and ten-frames to show that 10% of 40 = 40% of 10.

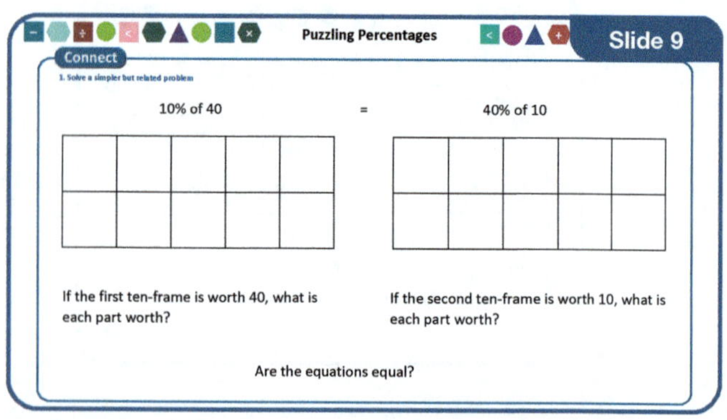

Encourage children to explain their reasoning.

- The first ten-frame represents 40 as the whole. 10% of 40 is 4 so the counter represents 4.
- The second frame represents 10 as the whole. Each counter represents 1. 10% is 1 so 40% is 4.
- Both frames have different wholes but the total of the counters in both equals 4.

2. Make a model / act it out

Additional resources required: base-10 blocks, bead strings

Explain to children that they can build models that show the commutative relationship between the equations and that this will allow them to provide appropriate proof. Emphasise that they should explain how their models show the equality in the equations each time.

Slide 10

Display slide 10. Ask children to replicate the image using their base-10 blocks (a hundreds-block with 7 ones-cubes placed on top.)

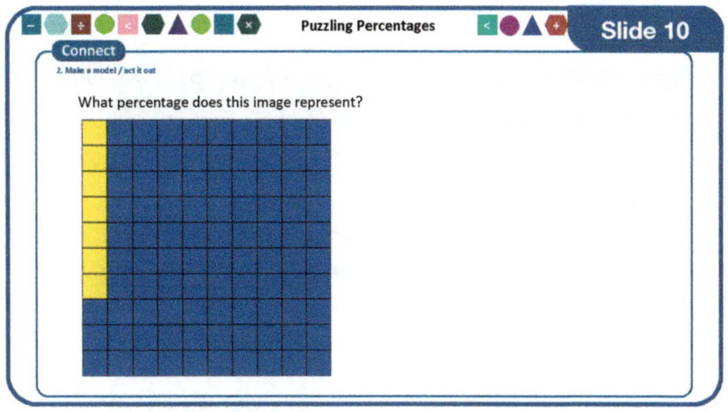

- *What percentage does this image represent?*

Agree that the hundreds-block represents 100% and that the 7 ones-cubes represent 7%.

- *If this image represents 7% of 100, what will 7% of 300 look like?*
- If the hundreds-block represents 300, then each individual cube within it represents 3 because 300 divided by 100 is 3. So 7% is 21 (7×3).

Slide 11

Display slide 11. Ask children to replicate the image with their base-10 blocks.

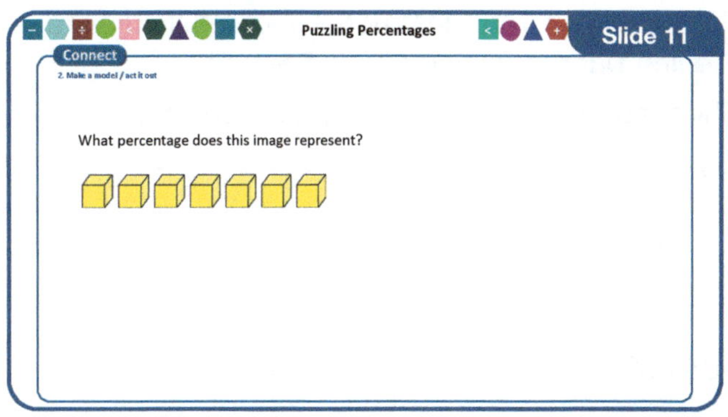

- *If this is 100% of 7, then what would 300% of 7 look like?*
- 300% is 3 lots of 100%. So 300% of 7 is equal to 21 (3×7).

Slide 12

Display slide 12. Agree that 7% of 300 is equal to 300% of 7.

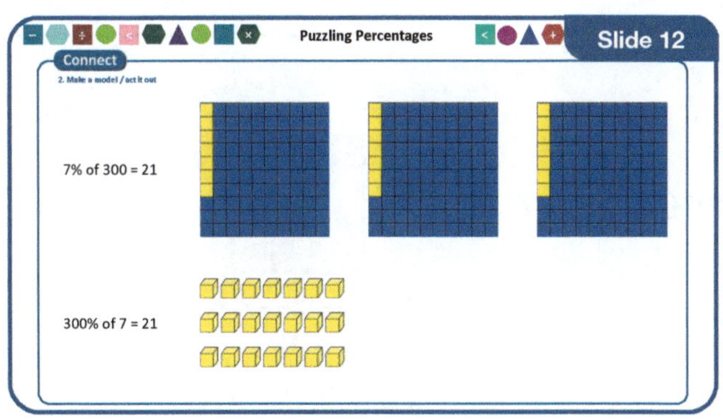

Encourage children to use other concrete resources to test other equations in the form a% of $b = b$% of a and look for further examples or counter examples. In this case, they will find no counter examples.

If children use bead strings as in the example below, check that they understand and explain that the bead strings represent different wholes.

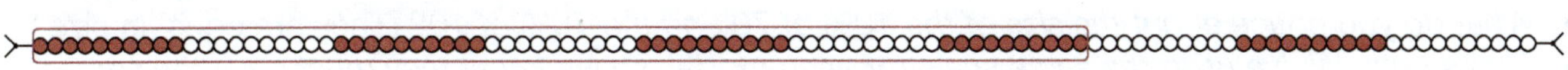

- I am proving that 10% of 70 equals 70% of 10. My two bead strings represent 70 and 10.
- If the whole is 70 then each bead represents 0·7. 10 beads of 0·7 equals 7. If the whole is 10 then each bead represents 0·01 with each group of 10 beads representing 0·1. 7 groups of 10 beads equals 7 or $0{\cdot}1 \times 70 = 7$.
- So 10% of 70 is equal to 70% of 10.

3. Draw a diagram

Additional resources required: rulers

Slide 13

Display slide 13 which shows two partially drawn bar models.

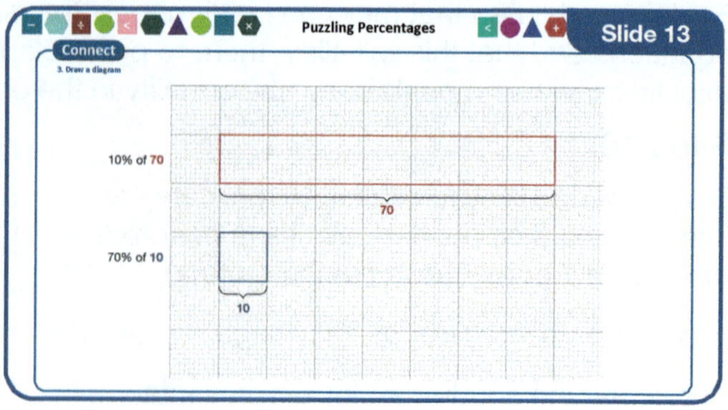

- *Can you help me to complete these bar models? We're trying to check whether 10% of 70 is equal to 70% of 10.*
- *In the first bar model, the whole is 70. How can we show 10%?*
- The bar is 70 squares wide. 10% of 70 is 7, so we could shade a part that is 7 squares wide.

⚠ Watch out for children who suggest that 10% is equal to 10 squares. Remind them that, in this case, 100% is 70, not 100, and therefore 10% is 7.

Following children's instructions, carefully draw a vertical line 7 squares in from the left-hand edge of the first bar model and shade the part.

Then repeat the process for the model showing 70% of 10.

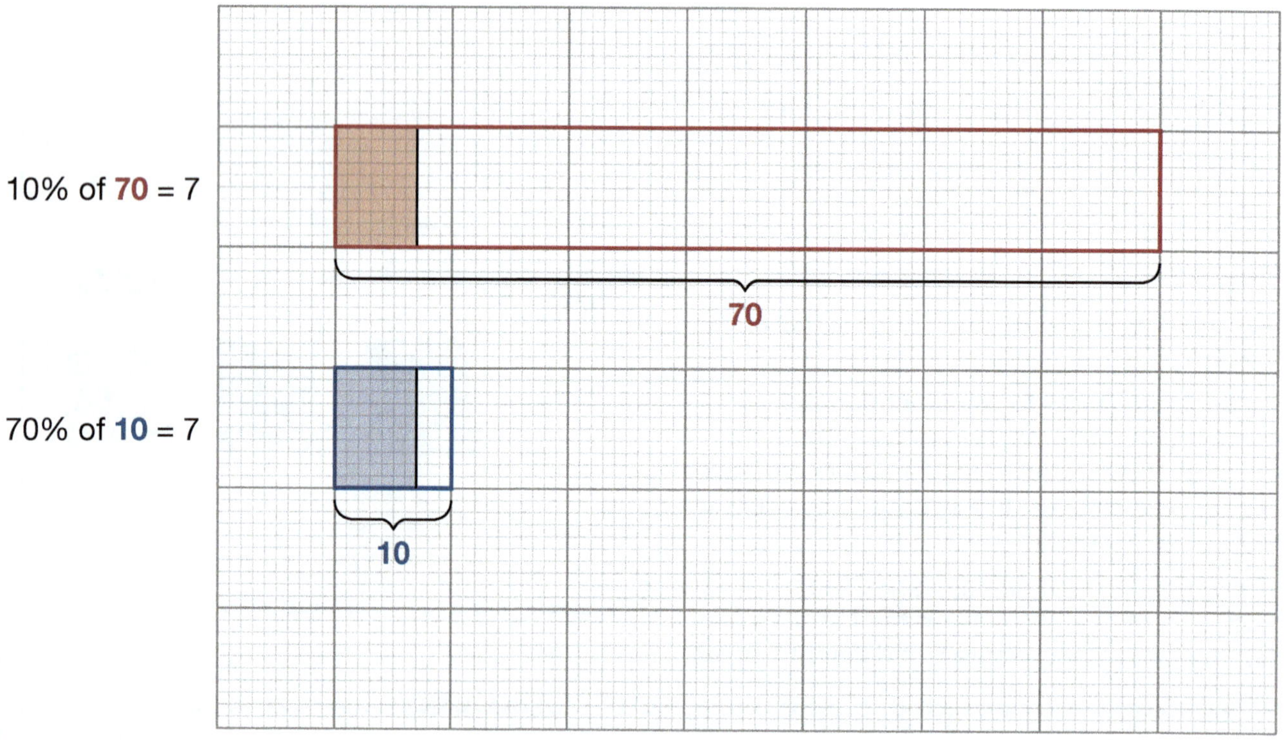

- *What do you notice about the size of the 10% of 70 and the 70% of 10?* (They are equal in size.)
- *That's right. We've used the same scale for each model, so we know they must have the same value.*

Slide 14

Print slide 14 (graph paper) and distribute to the children.

- *Now I'd like you to draw bar models to check whether 20% of 50 is equal to 50% of 20. I want you to use the graph paper to make sure your models are drawn to the same scale. You will need to be accurate.*

Support children with questioning as they work. For example:

- *What is the whole in each bar model?*
- *Why have you drawn one larger than the other?*
- *Have you shown the percentage required?*

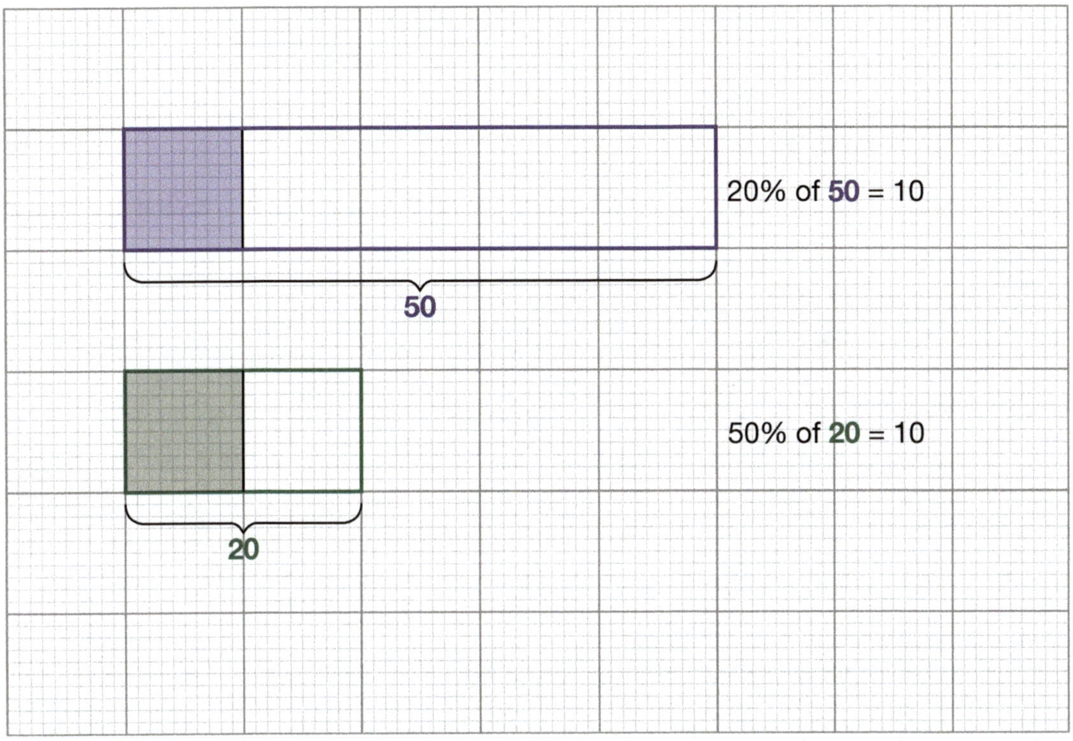

Give children more copies of slide 14 and ask them to write their own equations in the form of a% of $b = b$% of a and test them with bar models.

Refer to the **Collaborate** task of the **Pupil Resource**.

This task would work well with children working in pairs or groups.

Once children have explored the original equation 7% of 300 = 300% of 7, ask them to write different equations, with each child choosing one of the calculations to then prove that it is commutative.

Ask each child in the group to use a different type of concrete resource for their proof. They then share what they have discovered with their group, explaining how their model helps to show that the equation is true. Provide children with some sentence frames they can use if they need clarification from their partner. For example:

- I'm sorry, I didn't quite understand.
- Could you explain that in a different way?

Once all children have explained their equation, ask them to pass the equations clockwise and the concrete resources anticlockwise so that each child has a new equation and a different concrete resource. Ask them to prove the equation using their new resource. This puts the focus on the way the commutativity is proved rather than the answer to each side of the equation. Children could work in groups of eight, with pairs working together to prepare their proof and explanations.

Communicate

Refer to the **Communicate** task of the **Pupil Resource**.

This task provides an opportunity for the children to present their proof that these expressions create balanced equations. They could model one equation through a variety of representations, for example, proving that 7% of 300 is equal to 300% of 7 by using base-10 blocks, bead strings and ten-frames. Alternatively, they could choose one type of concrete resource and present proof for a range of equations. These presentations could be filmed or could be presented live to an audience.

Ask children to consider the following:

- *Would preparing a script / note cards be useful?*
- *How will you show your model clearly to your audience / the camera?*
- *How will you engage your audience?*

Create

Refer to the **Create** task of the **Pupil Resource**.

Slide 15

Display slide 15.

- *Were there any questions asked at the start of the task that you now want to investigate?*
- *Is there anything new that you are wondering now?*

Children can use the word bank to prompt further questions of their own.

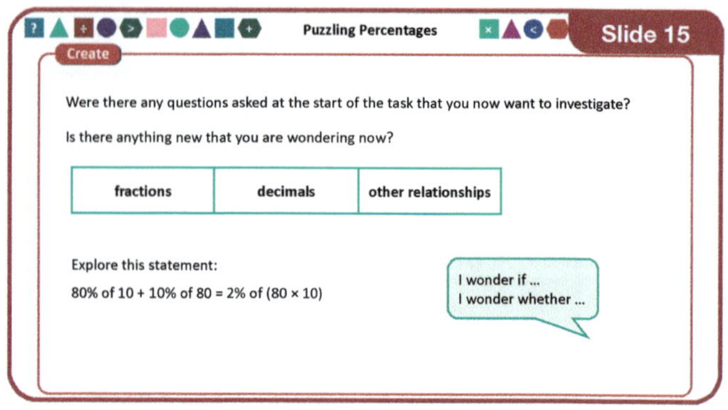

fractions	decimals	other relationships

- <u>I wonder</u> if this works for fractions and decimals.
- <u>I wonder</u> whether there are other relationships that I can find.

For example, children may suggest the following statement to investigate: 'I wonder if $\frac{1}{4}$ of 8 = 8 of $\frac{1}{4}$.' Here is an example of interlocking cubes used to investigate, and the reasoning involved.

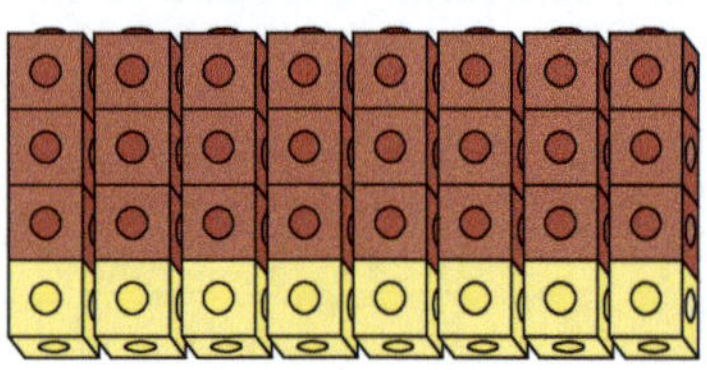

$\frac{1}{4}$ of 8 = 2

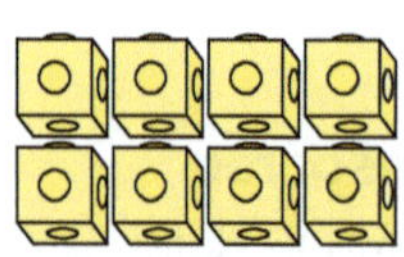

8 lots of $\frac{1}{4}$ = 2

- I know that a quarter of 8 equals 2 because 8 divided by 4 is 2.
- I have built 8 wholes. The red cubes represent three-quarters of each whole and the yellow cubes represent one-quarter. There are 8 yellow quarters. 8 quarters are equivalent to 2 wholes.
- Therefore $\frac{1}{4}$ of 8 is equal to 8 of $\frac{1}{4}$.

Provide children with the following statement to explore (this is on slide 15):

80% of 10 + 10% of 80 = 2% of (80 × 10).

Suggest that children break the equation into separate expressions and explore each separately.

- *Which expression is easiest to solve first?*
- *What is 2% of (80 × 10)? Is it 2% of 800?*
- *If this equals 16 and this expression is true, then 80% of 10 + 10% of 80 will equal 16. How can we test this?*

Refer to the Consider task of the **Pupil Resource**.

Encourage children to think about the strategies that they have used. The prompts below may help get children started.

Cognitive learning domain

- I think the representation that shows the equality clearest is …
- Some children may still struggle to understand the equality because … To help them I would …
- The best strategy I learned was … This changed my thinking because …

Affective domain

- I was influenced by … and as a result …
- I tried to influence others by …
- I was surprised about … because …

Metacognitive domain

- I found … unhelpful
- I used … to help me. It was helpful because …
- To work successfully as part of a group, next time I will …

Answers

The given equation is true with both expressions being equal to 21.

Answers to children's chosen equations will vary.

The Red Rectangle

Year 5

Introduction

The Red Rectangle is a challenging puzzle that requires children to apply their knowledge of area, multiples and factors.

To find the area of the red striped rectangle, children will need to find an efficient starting point and think logically. Rather than attempting to test all possible combinations, children will need to work systematically by using common factors and identifying dimensions that will fit and disregarding those that will not.

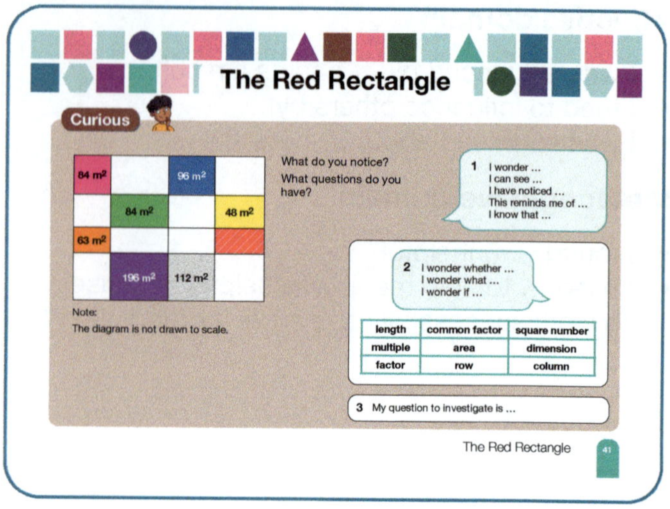

Children will also have the opportunity to rehearse their multiplication and division strategies in order to identify factors of the given numbers and should make links between known factors and multiplicative laws in order to find all factors.

This task promotes greater-depth learning in a range of ways, supporting children to identify and articulate underlying mathematical structures. For example, in the **Connect** stage, greater depth is promoted by asking children to explore a range of different mathematical resources and models, choose appropriate resources for the task and explain their choice. They will also monitor their progress towards the solution, and adjust their strategy if needed.

This **Teacher Guide** is supported by photocopiable **Pupil Resources** and **Teacher's Presentation** PowerPoint slides for whole-class or group display.

National Curriculum

The Red Rectangle covers these Year 5 National Curriculum objectives.

Measurement

- Calculate and compare the area of rectangles (including squares), and including using standard units, square centimetres (cm^2) and square metres (m^2) and estimate the area of irregular shapes.

Multiplication and division

- Identify multiples and factors, including finding all factor pairs of a number, and common factors of two numbers.
- Multiply and divide numbers mentally drawing upon known facts.

The 6C cycle

The grid on the right shows the relative weighting given to each of the 6 Cs in **The Red Rectangle**.

Curious	★★
Connect	★★★
Collaborate	★★
Communicate	★★
Create	★★
Consider	★★

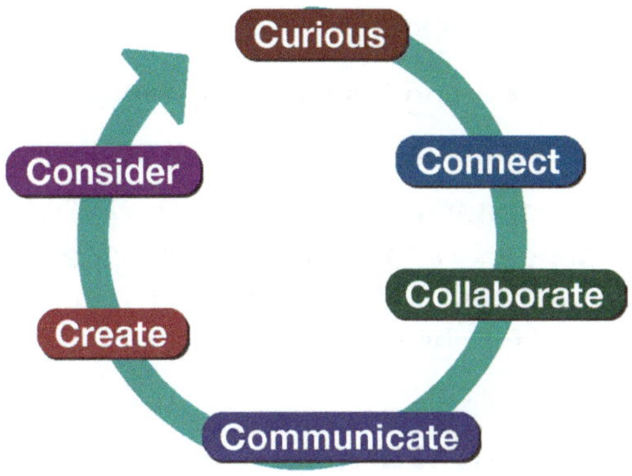

Problem-solving heuristics

The grid shows which problem-solving strategies children are likely to use when investigating the Connect tasks in **The Red Rectangle**. These strategies are further exemplified in the Connect stage below (pages 119–122).

Draw a diagram	✓
Make a table / spot a pattern	✓
Make a model / act it out	
Try then improve	✓
Create a systematic list	✓
Solve a simpler but related problem	✓
Work backwards	

Key mathematical ideas

The areas chosen have a number of factors and so children must use logical thinking to deduce the dimensions.

Some of the given areas go beyond 12×12 multiplication and division facts, so children will need to employ their knowledge of rules of divisibility and relationships between facts (such as doubling and halving or partitioning) to find all the factor pairs. For example, to work out the factors of 112:

- 1×112;
- 2×56 (using the knowledge that 2 must be a factor as 112 is an even number, and finding the matching factor by halving 112);
- 4×28 (using the rule that if the last two digits are divisible by 4 then the whole number will be divisible by 4; then using the doubling and halving rule); or using the known facts that both 100 and 12 are divisible by 4, then use partitioning.
- 8×14 (doubling and halving again);
- 16×7 (doubling and halving again).

Other factors that will be identified are:

48: 1×48, 2×42, 3×16, 4×12, 6×8

63: 1×63, 3×21, 7×9

84: 1×84, 2×42, 3×28, 4×21, 6×14, 7×12

96: 1×96, 2×48, 3×32, 4×24, 6×16, 8×12 (Children may use the rule that if the digit sum of a number is a multiple of 3, the whole number will be divisible by 3: the digit sum of 96 is 15.)

196: 1×196, 2×98, 4×49, 7×28, 14×14 (Children may recognise square numbers greater than 100, and then doubling and halving can be used to deduce that if $14 \times 14 = 196$ then 7×28 will also equal 196.)

Answers to the tasks are given in situ throughout the tasks and also on page 124.

Teaching guidance

If you are using the **Express Pathway**, please turn to pages 119–123.

Teaching guidance for the **Full 6C Pathway** is given below.

The **Teacher's Presentation** PowerPoint file has slides specifically mentioned in this teaching guidance. Some of the slides include notes (within the PowerPoint file) providing 'stumble support' suggestions, that may help you to support children in their problem-solving thinking and enable access for children who are less confident.

Awe and wonder

Slide 1

Display slide 1.

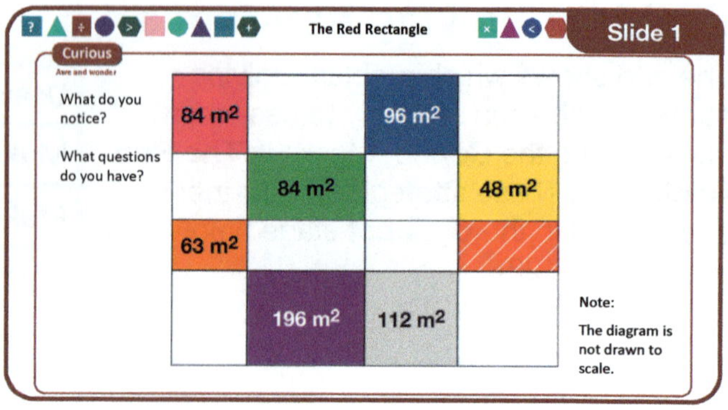

- I can see that the areas for some of the rectangles are shown, but some are not.
- I can see that six of the areas are even numbers, and one of the areas is an odd number.
- I have noticed that all of the areas are over 40 m².
- This reminds me of a multiplication square.
- I know that some of the rectangles have sides longer than 12 m, since $12 \times 12 = 144$.

⚠ A secure understanding of how area is calculated is essential for this task.

Develop and deepen

Encourage children to use the words in the word bank to develop and deepen their thoughts and questions.

length	**common factor**	**square number**
multiple	**area**	**dimension**
factor	**row**	**column**

Possible lines of enquiry:

- I wonder whether there is more than one possible solution.

Possible lines of enquiry linked to the word bank:

- I wonder what the dimensions of each coloured rectangle are.
- I wonder what the missing area is for the red striped rectangle.
- I wonder what would be the most efficient way of working out the missing lengths.
- I wonder what the dimensions of the large outer rectangle are.

Possible lines of enquiry that children may raise now, but should be held back until the **Create** task:

- I wonder if I could create my own missing-area puzzle.

Consider whether / how to collate and display all the suggested questions, and whether / how to steer children towards selecting an appropriately challenging question.

Pupil Resources for Full and Express Pathways

Distribute the **Pupil Resources** for **The Red Rectangle** to the children to help support them in their exploration of the different tasks. Use either of the **Curious Pupil Resources** depending on whether you are following the **Full Pathway** or the **Express Pathway**. The **Pupil Resources** for the remaining 5Cs are common to both pathways.

To provide additional flexibility in the classroom, each of the **Pupil Resources** is also provided in the **Teacher's Presentation** PowerPoint file (see slides 2–5), so you can display these at any time if you wish.

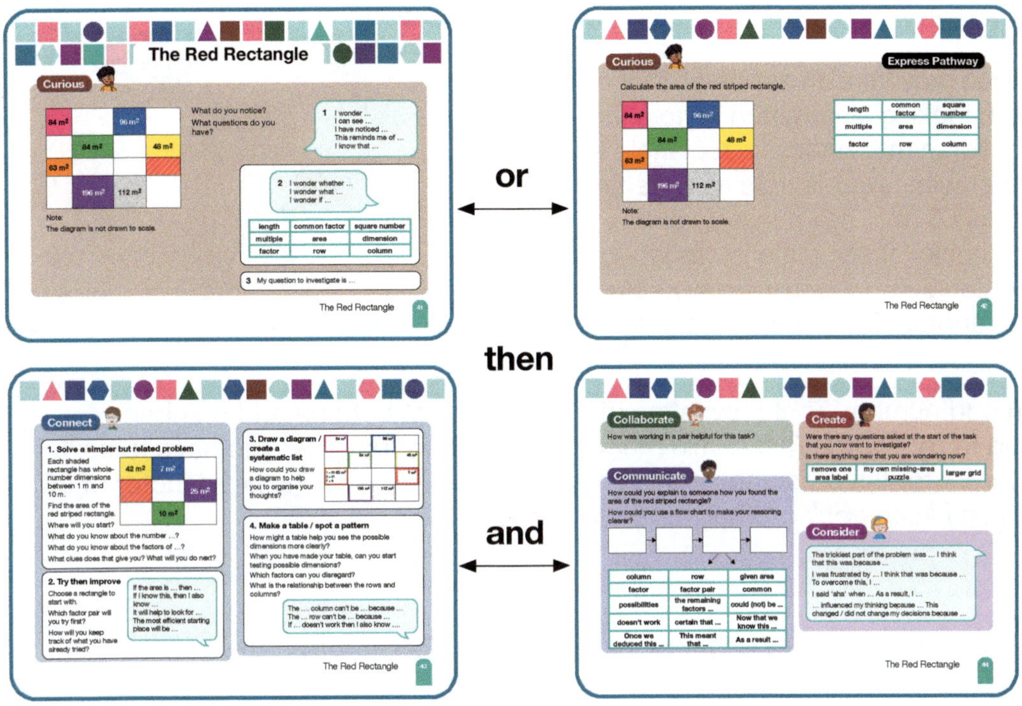

Express Pathway

If you are following the **Express Pathway**, spend a few minutes discussing the **Curious Pupil Resource** with the children. (This is slide 3, so you could display it if you wish.) Use some of the prompts on the previous page to help.

Four tasks involving different strategies are suggested for children to make connections.

1. Solve a simpler but related problem

Slide 6

Print slide 6 and distribute to children.

Encourage them to think about what they know about the numbers in this new diagram, to work out the area of the red striped rectangle.

- I know that 42 has lots of factors, so this might not be the best place to start.
- I know that 25 is a square number. It is 5 × 5. It also has factors of 1 and 25. So one of the dimensions of the red striped rectangle could be 5 m, 1 m or 25 m.

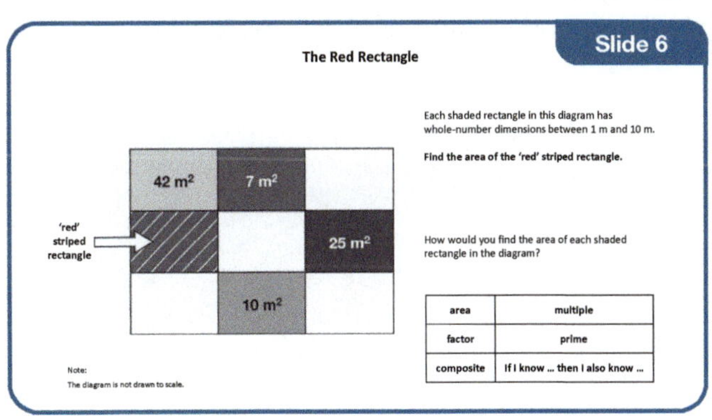

Connect

Children can then test to see which of these factors might work.

- It can't have dimensions of 1 m and 25 m, because the clue says that the largest dimension is 10 m. So it must be 5 m × 5 m.

Children could fill in the areas of additional rectangles based on what they discover.

2. Try then improve

Slide 7

Display slide 7.

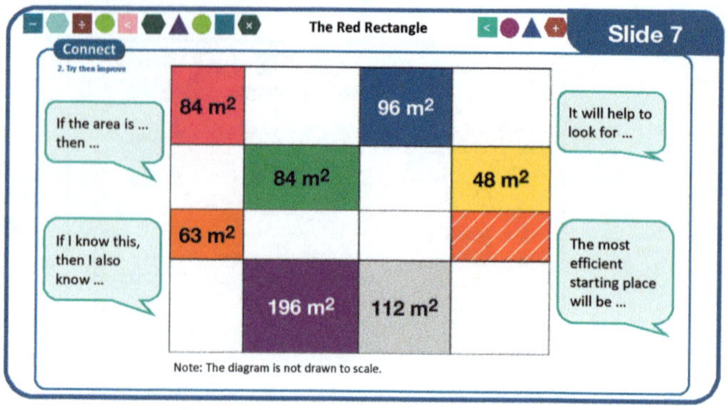

Some children may take an approach that does not start with looking at common factors. For example, children may choose a rectangle that has a product with fewer factors as a starting point, such as 63 m². They might try out a factor pair, check if those dimensions work for the other rectangles in the same row / column, and then try another factor pair if that does not work.

- 63 is an odd number. That means the dimensions must both be odd numbers.
- I'm going to try 1 × 63. That would mean the top-left rectangle would have to be 1 × 84. But that doesn't work for the other rectangle in the top row, because 84 is not a factor of 96.
- *How will you keep track of what you have already tried?*

3. Draw a diagram / create a systematic list

Slide 8

Print slide 8 and distribute to children.

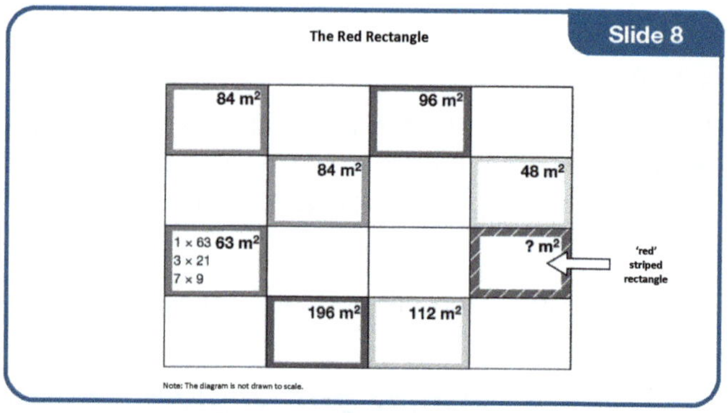

- *What do you think about what has been filled in here?*
- *What could you try next?*

Children could fill in the possible factor pairs for each area. Encourage children to list the factors in a systematic order to check that they have included all factor pairs. This will also support them in using strategies such as doubling and halving to check divisibility.

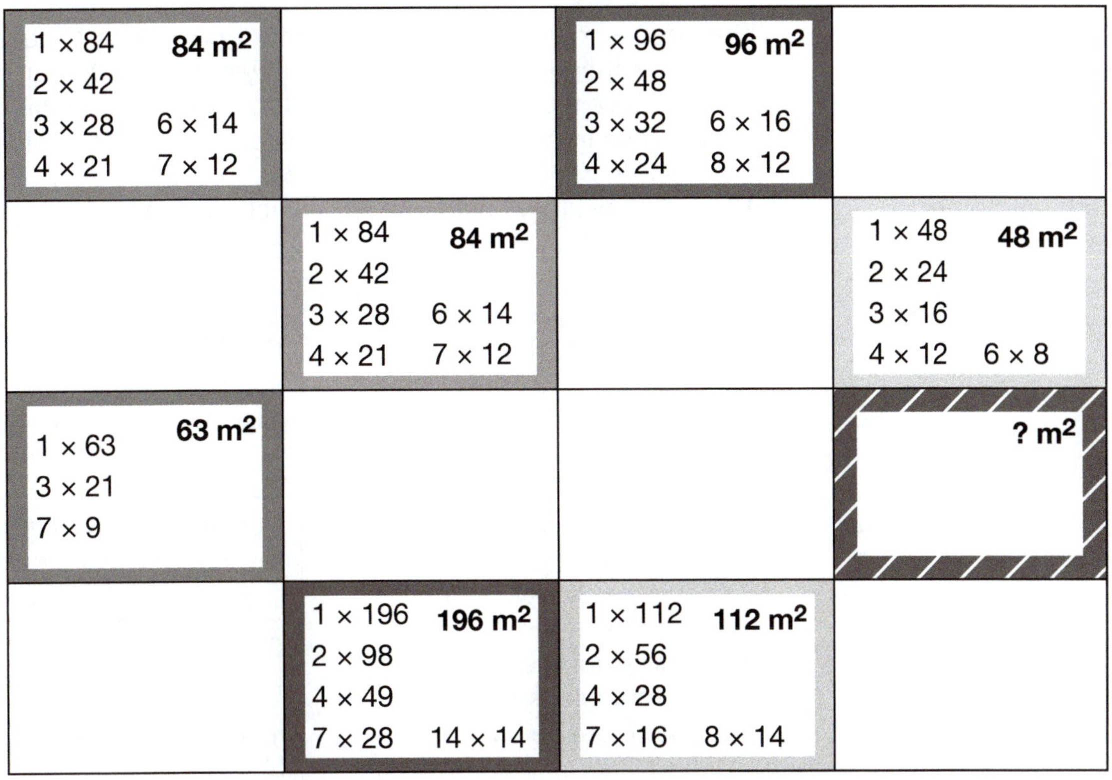

Children can then start to identify factors that will and will not work.

- The width of the first column must be 1, 3, 7 or 21. These are the only common factors in the two rectangles in this column.
- There are six common factors in the first row: 1, 2, 3, 4, 6 and 12. So the height of the first row could be any of those.

⚠ Watch for children who do not correctly identify all the factor pairs. Do they have sufficient strategies to do this?

Slide 9

Display slide 9.

- *Why might two copies of this diagram be useful?* (So we can compare the columns and the rows separately.)

You could give children a second print-out of slide 8, so they can create a copy. They could then use two different colours to highlight the common factors.

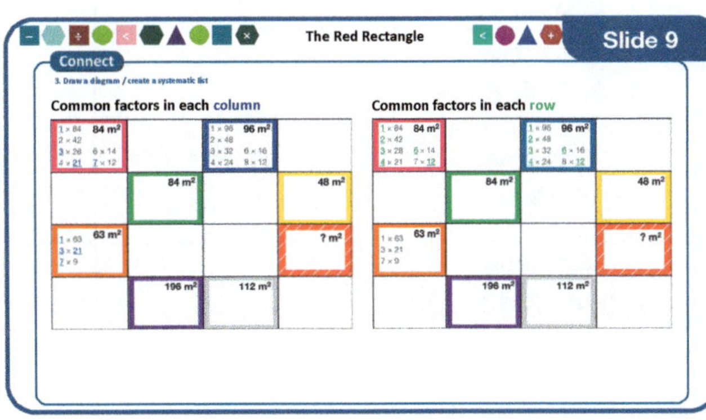

4. Make a table / spot a pattern

Slide 10

Display slide 10.

The common factors of each row and column could be transferred into a table. From here, children can identify factors that can be disregarded, and deduce the dimensions for the fourth column and third row.

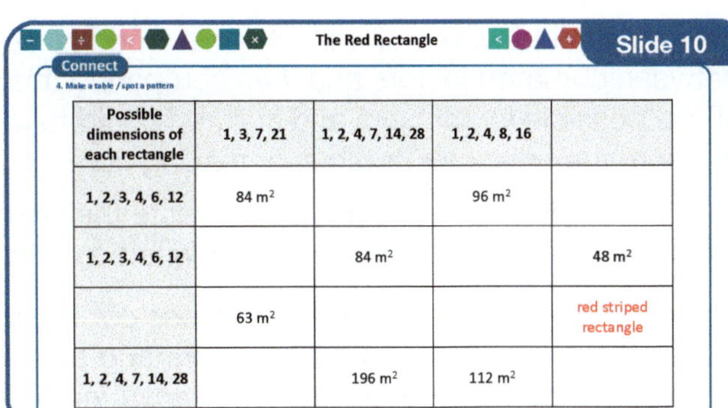

Possible dimensions of each rectangle	1, 3, 7, 21	1, 2, 4, 7, 14, 28	1, 2, 4, 8, 16	
1, 2, 3, 4, 6, 12	84 m²		96 m²	
1, 2, 3, 4, 6, 12		84 m²		48 m²
	63 m²			red striped rectangle
1, 2, 4, 7, 14, 28		196 m²	112 m²	

Connect

Row 1 cannot be 1, 2, 3 or 6. (Because those factors would need to pair with 84, 42, 28 or 14 to make 84, and these are not common factors of 63.)

→ That means row 1 must be either 4 or 12. (Because the only common factors for row 1 were 1, 2, 3, 4, 6 and 12, and I have ruled out 1, 2, 3 and 6.)

→ Row 1 cannot be 4, because that would need to pair with 24 to make 96, and 24 is not a possible dimension for column 3. So row 1 is 12, which means that column 3 is 8. (12 × 8 = 96)

→ If column 3 is 8, that means row 4 is 14. (Because 112 = 14 × 8.)

→ Now that I know that row 4 is 14, that means column 2 is also 14. (Because 196 = 14 × 14.)

→ If column 2 is 14, row 2 has to be 6. (Because 84 ÷ 14 = 6.)

→ So column 4 must be 8. (Because 48 = 6 × 8.)

Because I know that row 1 is 12, I also know that column 1 is 7. (Because 84 = 12 × 7.)

→ So row 3 must be 9. (Because I know that 7 is one of the dimensions for the 63 m² rectangle, so the other must be 9.)

→ I can now work out the area of the red striped rectangle: 9 m × 8 m = 72 m².

Possible dimensions of each rectangle	1st column 1, 3, **7**, 21	2nd column 1, 2, 4, 7, **14**, 28	3rd column 1, 2, 4, **8**, 16	4th column 16, 4, 8
1st row 1, 2, 3, 4, 6, 12	84 m² 12 × 7		96 m² 12 × 8	
2nd row 1, 2, 3, 4, 6, 12		84 m² 6 × 14		48 m² 6 × 8
3rd row 9	63 m² 9 × 7			red striped rectangle 9 × 8
4th row 1, 2, 4, 7, 14, 28		196 m² 14 × 14	112 m² 14 × 8	

Collaborate

Refer to the **Collaborate** task of the **Pupil Resource**.

This task would work well with children working independently at first to explore the connections between common factors and then comparing this with a partner. Pairs would work well when finding all the possible factor pairs and children could take turns to record their thinking either by crossing out from their diagram or table, or making notes.

Children could then work in pairs to create their explanation text, presentation or flow chart as detailed in the **Communicate** task or when creating their own puzzle as detailed in the **Create** task.

- *How did you decide together to cross out these factors?*
- *Did you both start from the same starting point?*
- *Has your partner made you realise anything new?*

Refer to the **Communicate** task of the **Pupil Resource**.

Slide 11

Display slide 11.

This task provides an opportunity for the children to create a written or verbal explanation detailing how they used the given areas to identify common factors, worked out which factors to disregard and worked logically to deduce each of the dimensions. This could also be presented as an explanation flow chart.

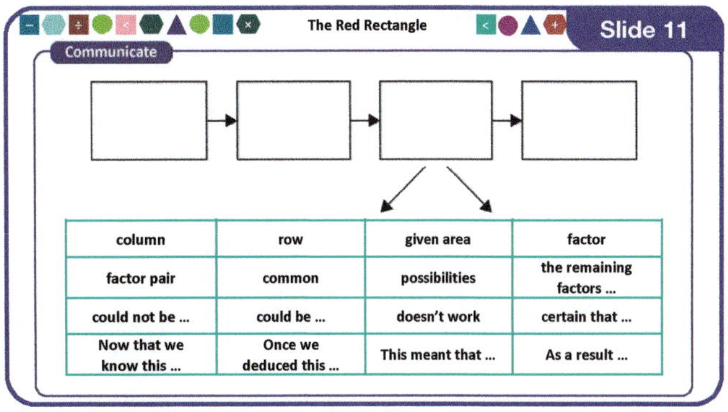

The blank flow chart and word bank provided could be useful to help children to structure and demonstrate their reasoning.

Refer to the **Create** task of the **Pupil Resource**.

- *Were there any questions asked at the start of the task that you now want to investigate?*
- *Is there anything new that you are wondering now?*

Children can use the word bank to prompt further questions of their own.

remove one area label	**my own missing-area puzzle**	**larger grid**

- I wonder what would happen if I removed one of the given areas.
- I wonder if I could create my own missing-area puzzle.
- I wonder if I could create a missing-area puzzle with a larger grid, for example, with 4 rows and 5 columns.

Slide 12

Display slide 12, if children choose to create their own puzzles.

Refer to the **Consider** task of the **Pupil Resource**.

Slide 13

Display slide 13.

Encourage children to think about the strategies that they have used. The prompts below may help get children started.

Cognitive learning domain

- The trickiest part of the problem was … I think that this was because …
- The new strategies or ideas that I considered during the task were … This was because …

Affective domain

- I was frustrated by … I think that was because … To overcome this, I …
- I said 'aha' when … As a result, I …
- When I found the area of the red striped rectangle I felt … I think that this was because …

Metacognitive domain

- When I got stuck or frustrated, I …
- … influenced my thinking because … This changed / did not change my decisions because …

Answers

The red striped rectangle has an area of 72 m² and dimensions of 9 m and 8 m.

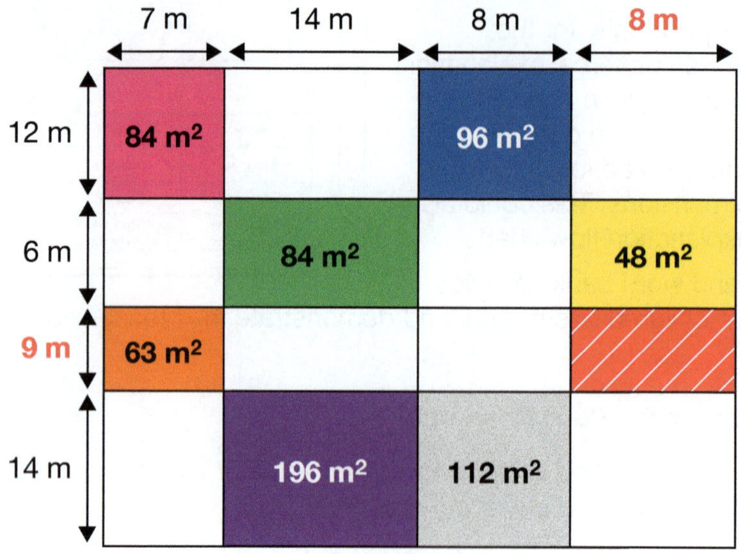

The large outer rectangle is 41 m by 37 m.

In the simpler puzzle in the first **Connect** task, the area of the red striped rectangle is 30 m² (5 m × 6 m).

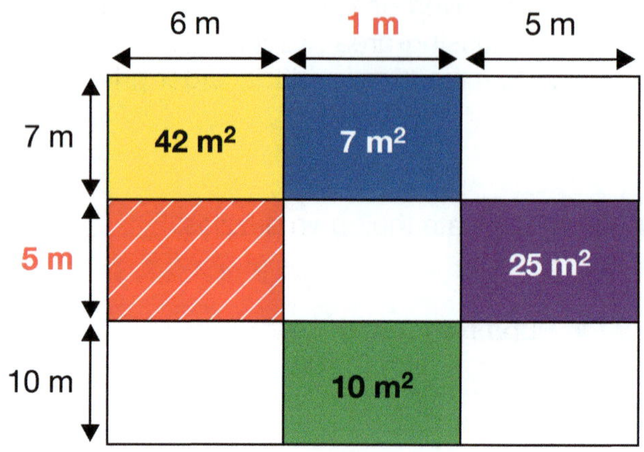

The Tale of Two Squares

Year 6

Introduction

The Tale of Two Squares is a step along the journey towards understanding and factorising algebraic expressions for the difference between two squares, which children will encounter when studying mathematics in later years. This task allows children to consolidate what they know about square numbers while considering patterns between them.

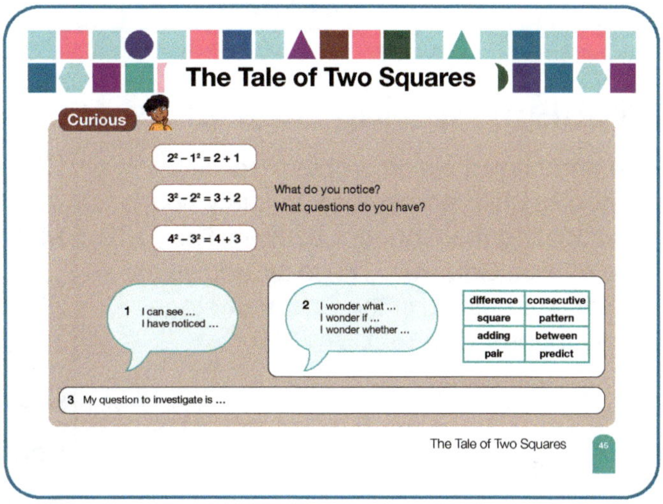

Through the use of their chosen heuristics, children will find ways to make these patterns easier to identify, and explain why the equations provided ($2^2 - 1^2 = 2 + 1$, $3^2 - 2^2 = 3 + 2$ and $4^2 - 3^2 = 4 + 3$) are true. Children will use accurate mathematical language alongside their models and diagrams to explain their findings.

Once children have investigated the initial problem, they may wish to explore whether the same patterns can be found with non-consecutive square numbers.

This task promotes greater-depth learning in a range of ways, supporting children to identify and articulate underlying mathematical structures. For example, in the **Curious** stage, greater depth is promoted by asking children to make rich mathematical contributions to discussions and suggest a variety of appropriate mathematical questions.

This **Teacher Guide** is supported by photocopiable **Pupil Resources** and **Teacher's Presentation** PowerPoint slides for whole-class or group display.

National Curriculum

The Tale of Two Squares covers these Year 6 National Curriculum objectives.

Measurement

- Recognise when it is possible to use formulae for area and volume of shapes.

Algebra

- *Pupils should be introduced to the use of symbols and letters to represent variables and unknowns in mathematical situations that they already understand, such as: equivalent expressions (for example, $a + b = b + a$); generalisations of number patterns; number puzzles (for example, what two numbers can add up to).* [non-statutory guidance]

The 6C cycle

The grid on the right shows the relative weighting given to each of the 6 Cs in **The Tale of Two Squares**.

Curious	★★★
Connect	★★★
Collaborate	★★
Communicate	★★★
Create	★★
Consider	★

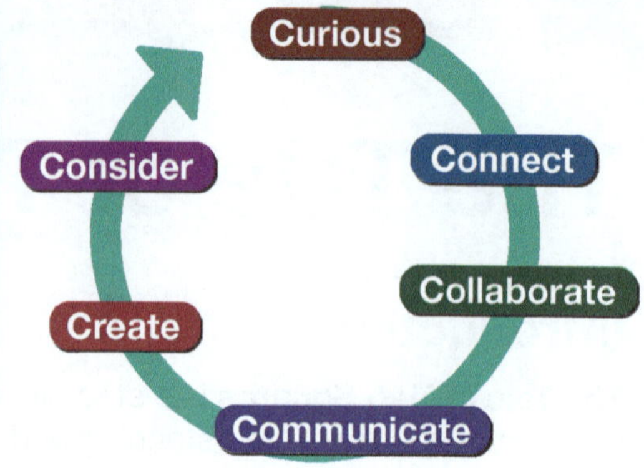

Problem-solving heuristics

The grid shows which problem-solving strategies children are likely to use when investigating the **Connect** tasks in **The Tale of Two Squares**. These strategies are further exemplified in the **Connect** stage below (pages 129–132).

Draw a diagram	✓
Make a table / spot a pattern	✓
Make a model / act it out	✓
Try then improve	
Create a systematic list	
Solve a simpler but related problem	
Work backwards	

Key mathematical ideas

Children will discover that, when finding the difference between the squares of two consecutive numbers, the value is the same as the sum of the numbers. For example:

$\mathbf{2^2 - 1^2 = 2 + 1}$

$\mathbf{2^2 - 1^2} = 4 - 1 = 3 = \mathbf{2 + 1}$

Drawing diagrams will help children to see why this happens.

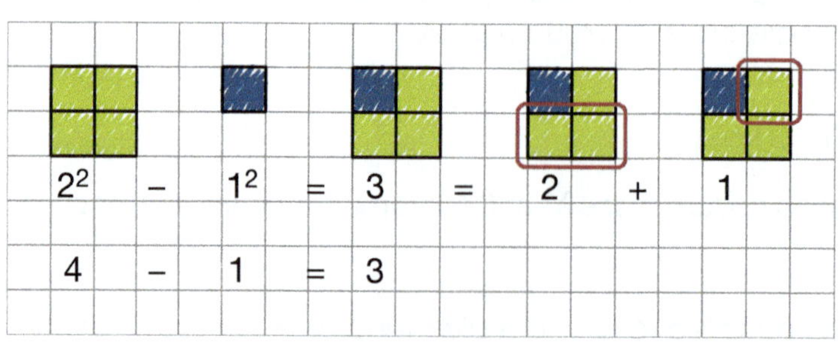

When the smaller square is layered on top of the larger square, the difference of 3 becomes clear.

For any pair of consecutive numbers, the following algebraic equation could be used:

$(n + 1)^2 - n^2 = (n + 1) + n$

The mathematical structure of the larger square, $(n + 1)^2$, could be shown using a diagram like this:

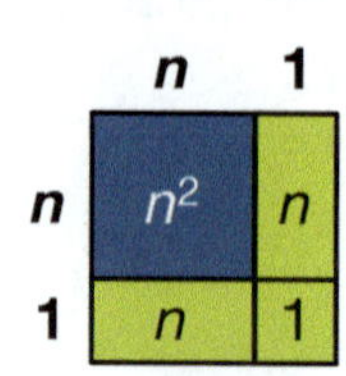

The diagram shows that

$(n+1)(n+1) = n^2 + n + n + 1 = n^2 + 2n + 1$

We could then show the difference between the larger square and the smaller square as follows:

$(n^2 + 2n + 1) - n^2 = 2n + 1$

Children are of course not expected to represent their findings in this form, but instead should look for patterns between the square numbers that they are exploring.

Answers to the tasks are given on page 134.

Teaching guidance

If you are using the **Express Pathway**, please turn to pages 128–133.

Teaching guidance for the **Full 6C Pathway** is given below.

The **Teacher's Presentation** PowerPoint file has slides specifically mentioned in this teaching guidance. Some of the slides include notes (within the PowerPoint file) providing 'stumble support' suggestions, that may help you to support children in their problem-solving thinking and enable access for children who are less confident.

Awe and wonder

Slide 1

Display slide 1.

Ask children to look at the three equations. Ask them to describe what they notice.

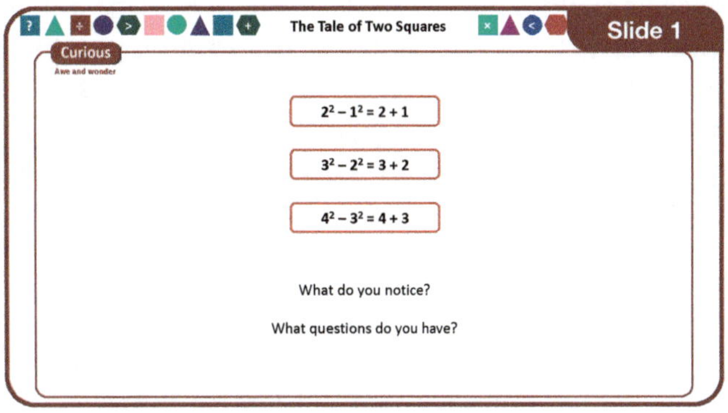

- I can see that the 'squared' symbol is used.
- I have noticed that there are number sentences either side of the equals sign.
- I have noticed that the numbers within each equation are consecutive.
- I have noticed that the numbers are the same on each side of the equation, but that on the addition side, the squared symbol is not used.
- I can see that the three equations are in a systematic order, with the numbers going up by 1 each time.
- I have noticed that the value of the three equations are odd numbers: 3, 5, 7.

Develop and deepen

Encourage children to use the words in the word bank to develop and deepen their thoughts and questions.

difference	**consecutive**	**square**	**pair**
pattern	**adding**	**between**	**predict**

Possible lines of enquiry:

- I wonder if the three equations are correct.

Possible lines of enquiry linked to the word bank:

- I wonder whether this pattern continues.
- I wonder if I could predict the difference between 100^2 and 101^2.
- I wonder if you can find the difference between squares of any pair of consecutive numbers by adding the numbers together.

Possible lines of enquiry that children may raise now, but should be held back until the **Create** task:

- I wonder if only odd numbers can be made like this.
- I wonder if the same thing happens when the numbers have a difference of 2 (for example, $4^2 - 2^2$).
- I wonder if the same thing happens with other non-consecutive numbers (for example, $5^2 - 1^2$).
- I wonder if all numbers can be made by finding the difference between two square numbers.
- I wonder if this would happen with cube numbers as well.

Consider whether / how to collate and display all the suggested questions, and whether / how to steer children towards selecting an appropriately challenging question.

Pupil Resources for Full and Express Pathways

Distribute the **Pupil Resources** for **The Tale of Two Squares** to the children to help support them in their exploration of the different tasks. Use either of the **Curious Pupil Resources** depending on whether you are following the **Full Pathway** or the **Express Pathway**. The **Pupil Resources** for the remaining 5Cs are common to both pathways.

To provide additional flexibility in the classroom, each of the **Pupil Resources** is also provided in the **Teacher's Presentation** PowerPoint file (see slides 2–5), so you can display these at any time if you wish.

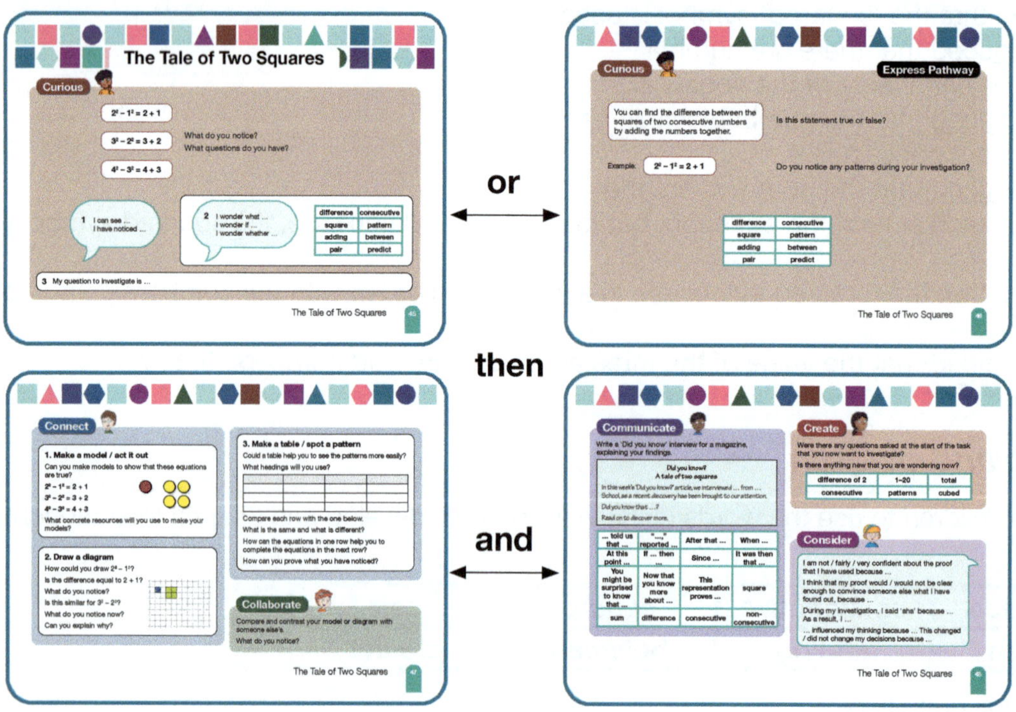

Express Pathway

If you are following the **Express Pathway**, spend a few minutes discussing the **Curious Pupil Resource** with the children. (This is slide 3, so you could display it if you wish.) Use some of the prompts on the previous page to help.

Three tasks involving different strategies are suggested for children to make connections.

1. Make a model / act it out

Additional resources required: cubes, counters or square sticky notes

Slide 6

Display slide 6.

Ask children to build models to represent 1^2 and 2^2, using different coloured cubes, counters or sticky notes to represent each number.

- *How can you use your models to clearly show that the difference between 1 squared and 2 squared is 3?*

⚠ If children use cubes, check that they do not confuse area with volume.

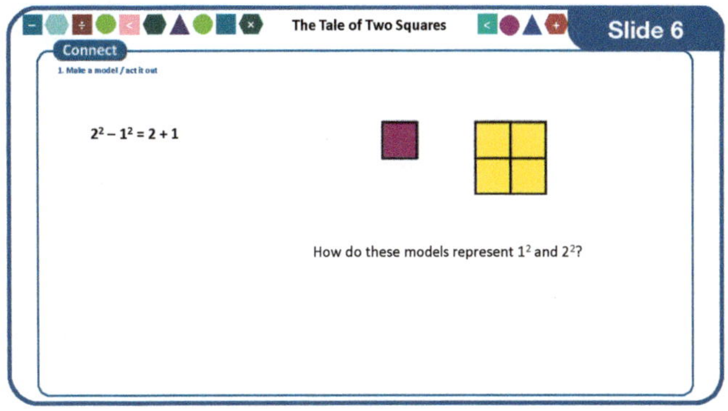

Slide 7

Display slide 7.

If needed, encourage children to place their smaller square on top of their larger square (or replace counters or sticky notes).

- *Where can we see the '3'?*

Children should then explore the other two equations from **Pupil Resource** page 45 ($3^2 - 2^2$ and $4^2 - 3^2$) by making similar models.

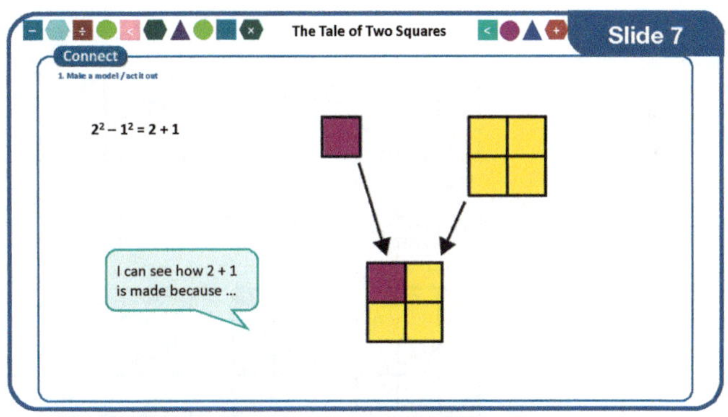

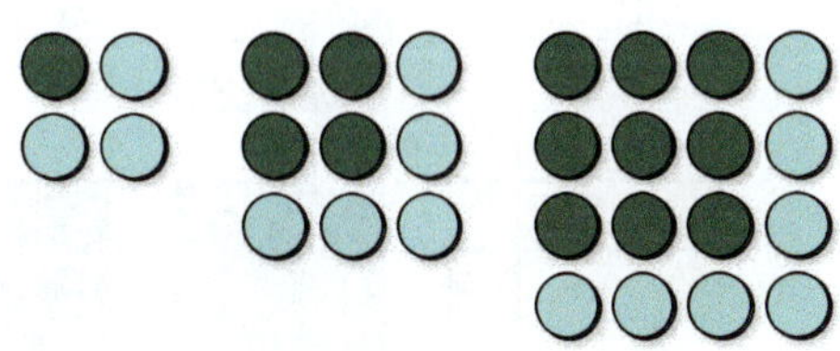

In both sets of models pictured here, the difference between the squares is clear to see. Children will also see why the total of the two consecutive numbers is equal to the difference between the squares.

$2 + 3 = 5$

$3^2 - 2^2 = 9 - 4 = 5$

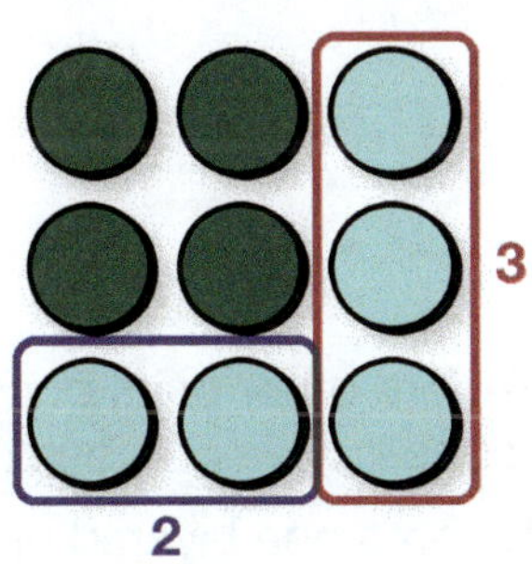

- I can see how the $2 + 3$ is made. The smaller square number (4) has 2 blue counters to the right, and 2 blue counters below, which is 4 counters in total. But to make a square, an extra counter needs to be added. So this is $2 + 2 + 1$, or $2 + 3$, or $3 + 2$.
- I could also write this as $2 \times 2 + 1$.

2. Draw a diagram

Additional resources required: squared paper, coloured pencils

Slides 8 and 9

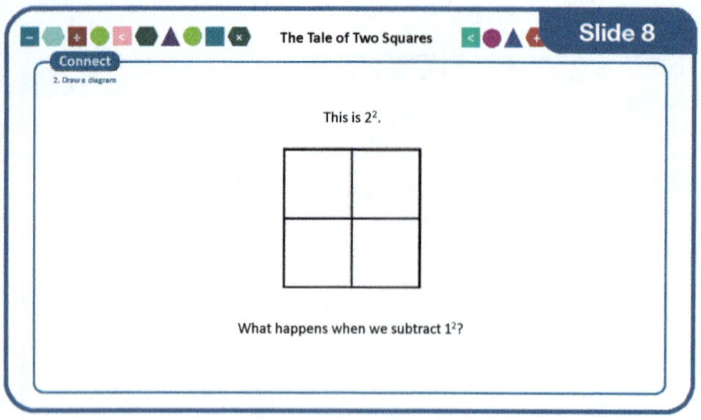

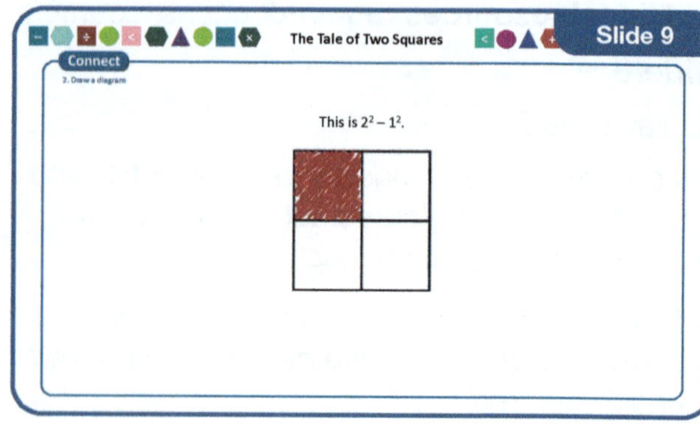

Display slide 8 and then slide 9 if needed.

Ask children to use squared paper to draw diagrams to show what is happening in the three given equations.

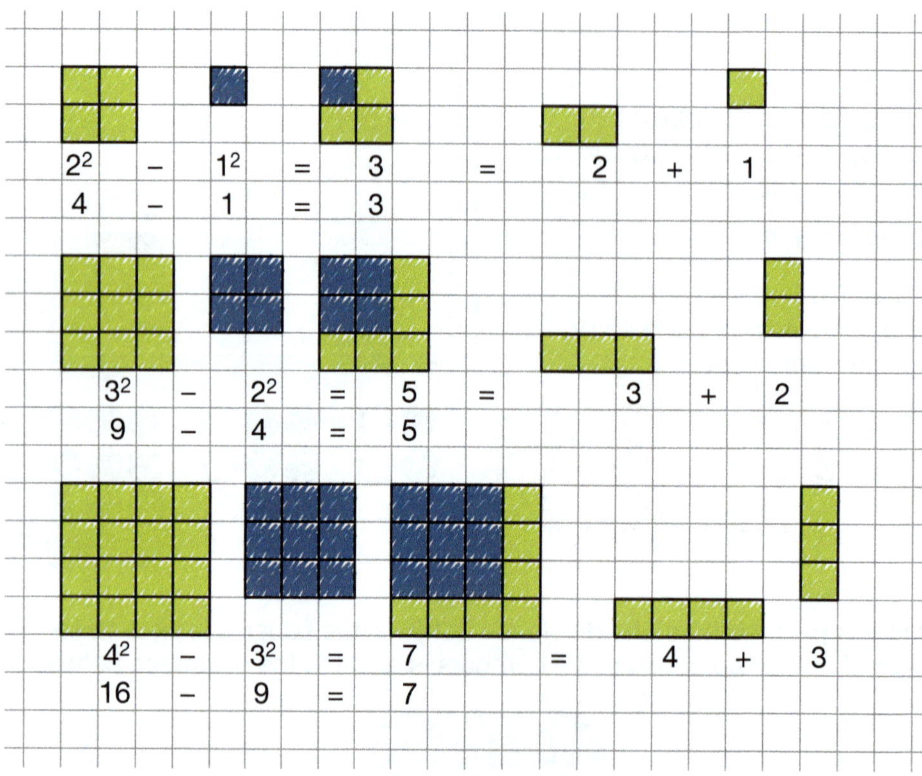

Diagrams showing that the rule works for pairs of consecutive numbers

They can then use the same method to explore other pairs of consecutive numbers. If they wish, they could then move on to investigating non-consecutive numbers.

As with making models, by drawing the squares one layered on top of the other, it is clear to see when the rule works (when the two numbers are consecutive) and when it does not (when not consecutive). Once children discover that it does not work when the numbers are not consecutive, this may lead them to exploring other relationships and patterns.

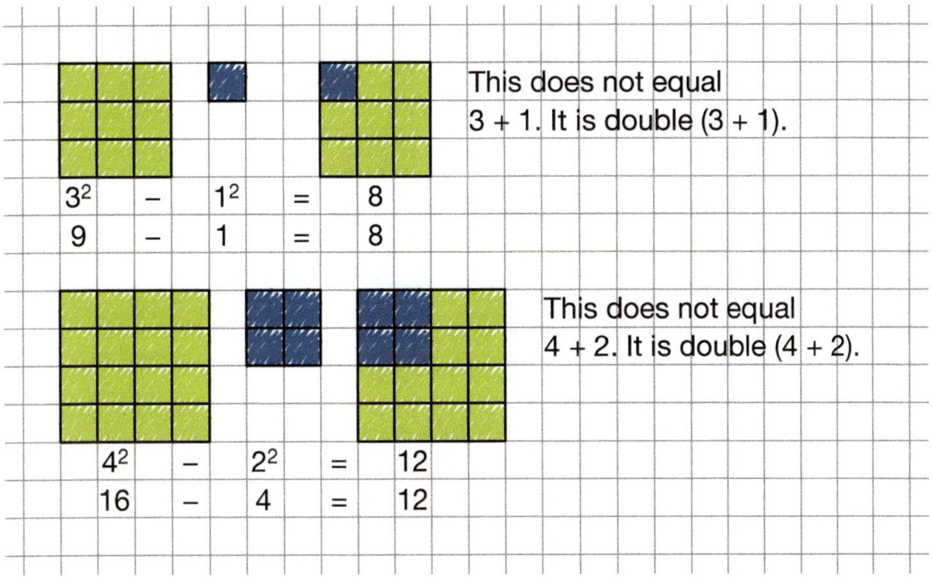

Diagrams showing that the rule does not work for pairs of non-consecutive numbers, but that another rule might apply when the two numbers have a difference of 2

3. Make a table / spot a pattern

Slide 10

Display slide 10.

- *What patterns can you see in your models or diagrams?*
- *Could a table help you to see the patterns more easily?*
- *What will be useful headings for the table?*

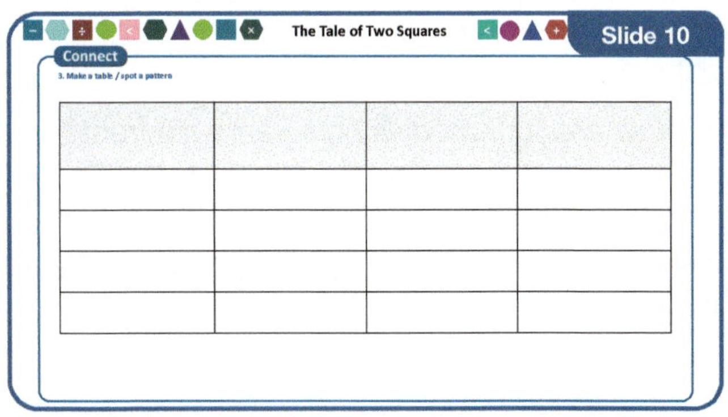

Give children time to make their own table and fill in the details for $2^2 - 1^2$ in the first row.

Slide 11

Display slide 11.

- *What is useful about these headings?*

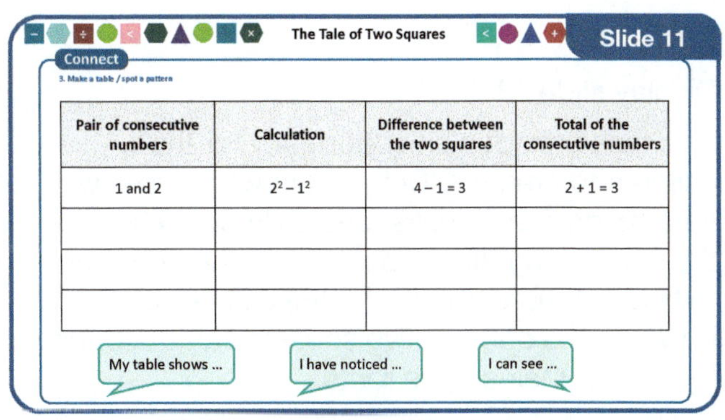

Fill in the second and third rows when children have explored both.

- *Can you fill in the row for 4 and 5 now?*

Connect

Pair of consecutive numbers	Calculation	Difference between the two squares	Total of the consecutive numbers
1 and 2	$2^2 - 1^2$	$4 - 1 = 3$	$2 + 1 = 3$
2 and 3	$3^2 - 2^2$	$9 - 4 = 5$	$3 + 2 = 5$
3 and 4	$4^2 - 3^2$	$16 - 9 = 7$	$4 + 3 = 7$
4 and 5	$5^2 - 4^2$	$25 - 16 = 9$	$5 + 4 = 9$

- *Does the table make it easier to spot patterns?*
- *How can you prove what you have noticed?*
- My table shows that the difference between the consecutive square numbers is the same as adding the two numbers together.
- I have noticed that all of the differences and totals are odd numbers.
- *Can the reason for the pattern be proved using your models or diagrams?*

Collaborate

Refer to the **Collaborate** task of the **Pupil Resource**.

Slide 12

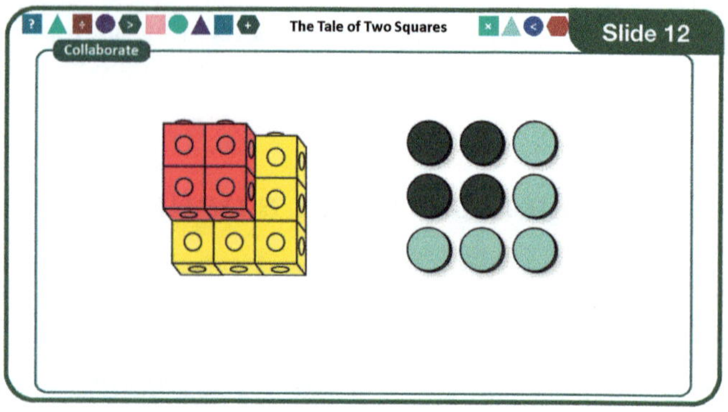

Display slide 12.

This task could effectively start with children working independently and then comparing with a partner what they have found out, with a focus on comparing their models and diagrams, and how these expose the reasons for the pattern.

- *How are your models / diagrams different?*
- *In what ways are they the same as yours?*
- *Whose models / diagrams do you think makes the pattern clearer?*

Children could then work in pairs to create their magazine article as detailed in the **Communicate** task.

Communicate

Refer to the **Communicate** task of the **Pupil Resource**.

Slide 13

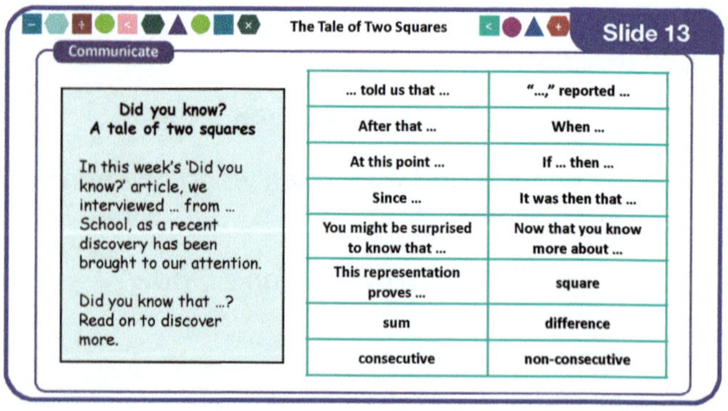

Display slide 13.

This task provides an opportunity for the children to create a 'Did you know?' interview for a magazine. In pairs, one child could take on the role of journalist and interview their partner to find out more about why the difference between the consecutive squares can be found by adding their square roots.

Children should be encouraged to rehearse their use of direct and indirect speech in their article, using accurate punctuation. Children should also provide appropriate models of proof.

Refer to the **Create** task of the **Pupil Resource**.

- *Were there any questions asked at the start of the task that you now want to investigate?*
- *Is there anything new that you are wondering now?*

Children can use the word bank to prompt further questions of their own.

difference of 2	**1–20**	**total**
consecutive	**patterns**	**cubed**

- I wonder if only odd numbers can be made like this.
- I wonder if the same thing happens when the numbers have a difference of 2 (for example, $4^2 - 2^2$).
- I wonder if the same thing happens with other non-consecutive numbers (for example, $5^2 - 1^2$).
- I wonder if all numbers can be made by finding the difference between two square numbers.
- I wonder if there is a pattern in the totals of consecutive square numbers.
- I wonder if this would happen with cube numbers as well.

Slide 14

If children choose to investigate whether they can make the numbers 1 to 20 by finding the difference between two square numbers, a table would be a helpful way for them to present their thinking and their findings, as shown on slide 14.

Number	Difference between two square numbers	Can it be done?
1	$1^2 - 0^2 = 1$	Yes
2	$2^2 - 1^2 = 3$ $2^2 - 0^2 = 4$ $3^2 - 2^2 = 5$	No
3	$2^2 - 1^2 = 3$	Yes
4	$2^2 - 0^2 = 4$	Yes

Try then improve could be used as an additional strategy.

Do children recognise that for numbers that are already square, such as 1 and 4, they can be found by subtracting 0^2 from them?

Children may also notice that multiples of 4 can be found by squaring numbers that have a difference of 2. For example, 8 can be written as $3^2 - 1^2$, 12 as $4^2 - 2^2$, and 16 as $5^2 - 3^2$.

Refer to the **Consider** task of the **Pupil Resource**.

Slide 15

Display slide 15.

Encourage children to think about the strategies that they have used. The prompts below may help get children started.

Cognitive learning domain

- I am not / fairly / very confident about the proof that I have used because …
- I think that my proof would / would not be clear enough to convince someone else what I have found out, because …
- I think that someone could argue against my proof because …

Affective domain

- The enjoyable part of my learning was … I enjoyed this element because ...
- During my investigation, I said 'aha' because … As a result, I …
- I was surprised that … because …

Metacognitive domain

- Some representations are more helpful than others because …
- I interacted with others in the class by … This was successful because …
- … influenced my thinking because … This changed / did not change my decisions because …

Answers

Consecutive numbers:

n is ...	$(n+1)^2 - n^2 = (n+1) + n$	Difference between the squares	2n + 1
0	$1^2 - 0^2 = 1 + 0$	$1 - 0 = 1$	$0 + 1 = 1$
1	$2^2 - 1^2 = 2 + 1$	$4 - 1 = 3$	$2 + 1 = 3$
2	$3^2 - 2^2 = 3 + 2$	$9 - 4 = 5$	$4 + 1 = 5$
3	$4^2 - 3^2 = 4 + 3$	$16 - 9 = 7$	$6 + 1 = 7$
4	$5^2 - 4^2 = 5 + 4$	$25 - 16 = 9$	$8 + 1 = 9$
5	$6^2 - 5^2 = 6 + 5$	$36 - 25 = 11$	$10 + 1 = 11$
10	$11^2 - 10^2 = 11 + 10$	$121 - 100 = 21$	$20 + 1 = 21$
100	$101^2 - 100^2 = 101 + 100$	$10\,201 - 10\,000 = 201$	$200 + 1 = 201$

Numbers with a difference of 2:

The difference between the squares is equal to double the sum of the numbers.

n is ...	$(n+2)^2 - n^2 = 2((n+2) + n)$	Difference between the squares	4n + 4
0	$2^2 - 0^2 = 2 \times (2 + 0)$	$4 - 0 = 4$	$0 + 4 = 4$
1	$3^2 - 1^2 = 2 \times (3 + 1)$	$9 - 1 = 8$	$4 + 4 = 8$
2	$4^2 - 2^2 = 2 \times (4 + 2)$	$16 - 4 = 12$	$8 + 4 = 12$
3	$5^2 - 3^2 = 2 \times (5 + 3)$	$25 - 9 = 16$	$12 + 4 = 16$
4	$6^2 - 4^2 = 2 \times (6 + 4)$	$36 - 16 = 20$	$16 + 4 = 20$
5	$7^2 - 5^2 = 2 \times (7 + 5)$	$49 - 25 = 24$	$20 + 4 = 24$
10	$12^2 - 10^2 = 2 \times (12 + 10)$	$144 - 100 = 44$	$40 + 4 = 44$
100	$102^2 - 100^2 = 2 \times (102 + 100)$	$10\,404 - 10\,000 = 404$	$400 + 4 = 404$

Which of the numbers 1–20 can be made by finding the difference between two squares?

Number	Can it be done?	Difference between two square numbers
1	Yes	$1^2 - 0^2$
2	No	-
3	Yes	$2^2 - 1^2$
4	Yes	$2^2 - 0^2$
5	Yes	$3^2 - 2^2$
6	No	-
7	Yes	$4^2 - 3^2$
8	Yes	$3^2 - 1^2$
9	Yes	$5^2 - 4^2$ or $3^2 - 0^2$
10	No	-

Number	Can it be done?	Difference between two square numbers
11	Yes	$6^2 - 5^2$
12	Yes	$4^2 - 2^2$
13	Yes	$7^2 - 6^2$
14	No	-
15	Yes	$8^2 - 7^2$ or $4^2 - 1^2$
16	Yes	$5^2 - 3^2$ or $4^2 - 0^2$
17	Yes	$9^2 - 8^2$
18	No	-
19	Yes	$10^2 - 9^2$
20	Yes	$6^2 - 4^2$

Odd numbers can be made by finding the difference between the squares of consecutive numbers.

Multiples of 4 can be made by finding the difference between the squares of numbers with a difference of 2.

Square numbers can be made by finding the difference between a square and the square of 0.

Talented Tessellation

Year 5

Introduction

Talented Tessellation challenges children's ideas and assumptions concerning perimeter and area.

They explore the mathematics through creating shapes that tessellate, so there is an opportunity to see and develop 'an appreciation of the beauty and power of mathematics' (Mathematics programmes of study: Key stages 1 and 2 National Curriculum in England, 2013 p.3).

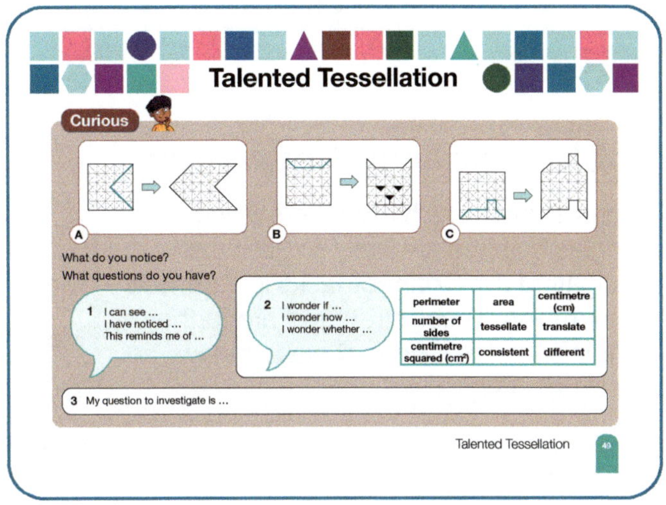

Communication provides the opportunity for children to see mathematics through a different perspective: as a tool employed by artists such as Escher, Bilney and Fathauer, and by different cultures throughout history (for example, Sumerians and Romans).

This task promotes greater-depth learning in a range of ways, supporting children to identify and articulate underlying mathematical structures. For example, in the **Communicate** stage, greater depth is promoted by asking children to develop clear, accurate and well-organised communication, with ideas connected and flowing from one to the next, and illustrated using diagrams or models for clarification as needed.

This **Teacher Guide** is supported by photocopiable **Pupil Resources** and **Teacher's Presentation** PowerPoint slides for whole-class or group display.

National Curriculum

Talented Tessellation covers these Year 5 National Curriculum objectives.

Geometry: position and direction

- Identify, describe and represent the position of a shape following a reflection or translation, using the appropriate language, and know that the shape has not changed.

Measurement

- Measure and calculate the perimeter of composite rectilinear shapes in centimetres and metres.
- Calculate and compare the area of rectangles (including squares), and including using standard units, square centimetres (cm^2) and square metres (m^2) and estimate the area of irregular shapes.

The 6C cycle

The grid on the right shows the relative weighting given to each of the 6 Cs in **Talented Tessellation**.

Curious	★★
Connect	★
Collaborate	★★★
Communicate	★★★
Create	★★
Consider	★★

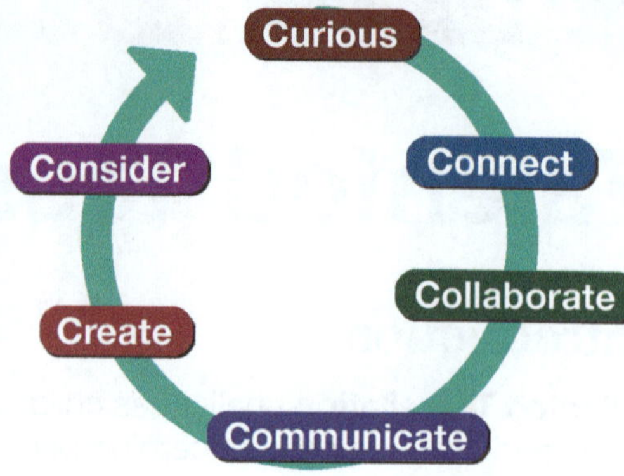

Problem-solving heuristics

The grid shows which problem-solving strategies children are likely to use when investigating the Connect tasks in **Talented Tessellations**. These strategies are further exemplified in the Connect stage below (pages 140–141).

Draw a diagram	
Make a table / spot a pattern	✓
Make a model / act it out	✓
Try then improve	✓
Create a systematic list	✓
Solve a simpler but related problem	
Work backwards	

Key mathematical ideas

Tessellating (or tiling) is when a surface is covered with flat shapes so that there are no gaps or overlaps. This task uses translation tessellation, in which the shape repeats by moving or sliding along. Through creating increasingly intricate tessellating patterns, children will discover that as they increase the number of sides, the perimeter increases. The area remains constant as nothing is being removed or added to the shape, it is only being rearranged.

In theory, there is no limit to how large the perimeter could get as it is always possible to add more sides by cutting and translating smaller and smaller sides.

For example, the shape on the next page starts as a simple tessellating shape, but as more parts are translated, the perimeter increases (but the area remains constant). The consistent translating of the cut-out parts means that the overall shape still tessellates.

For the perimeter lengths shown here, the length of the diagonal of a square has been approximated to 1·4 cm.

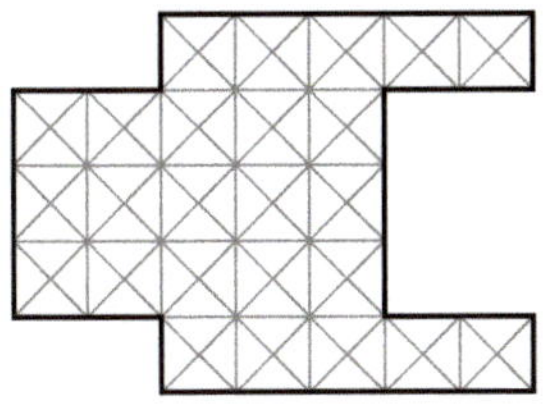

Number of sides: 12
Perimeter: 28 cm
Area: 25 cm^2

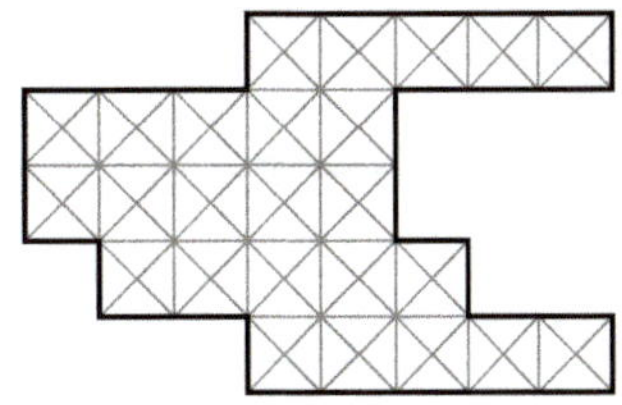

Number of sides: 16
Perimeter: 32 cm
Area: 25 cm^2

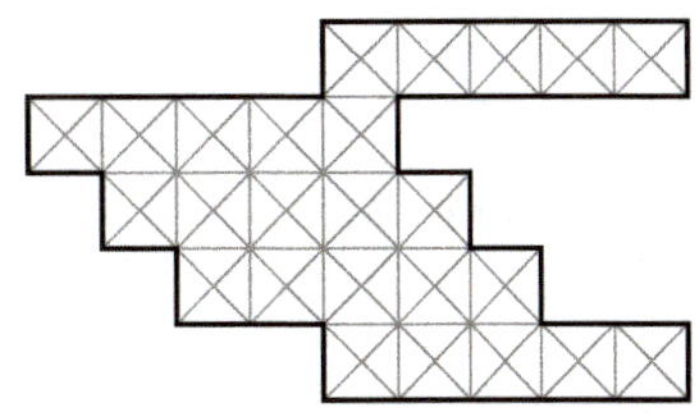

Number of sides: 20
Perimeter: 36 cm
Area: 25 cm^2

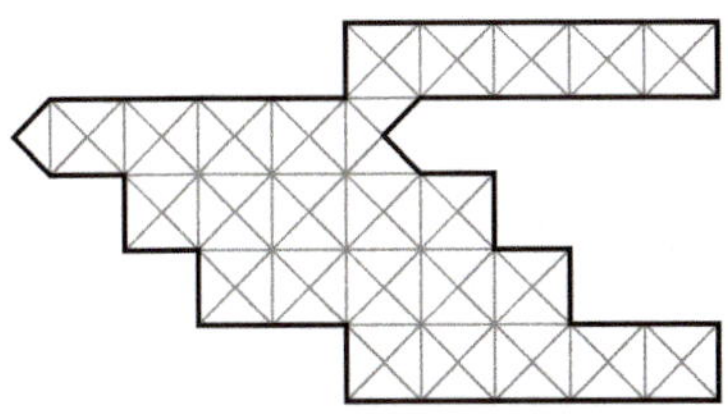

Number of sides: 22
Perimeter: 36·8 cm
Area: 25 cm^2

Answers are given in situ throughout the tasks because of the cumulative nature of the tasks.

Teaching guidance

If you are using the **Express Pathway**, please turn to pages 140–143.

Teaching guidance for the **Full 6C Pathway** is given below.

The **Teacher's Presentation** PowerPoint file has slides specifically mentioned in this teaching guidance. Some of the slides include notes (within the PowerPoint file) providing 'stumble support' suggestions, that may help you to support children in their problem-solving thinking and enable access for children who are less confident.

Awe and wonder

Additional resources required: scissors, sticky tape, plain paper

Slides 1 and 2

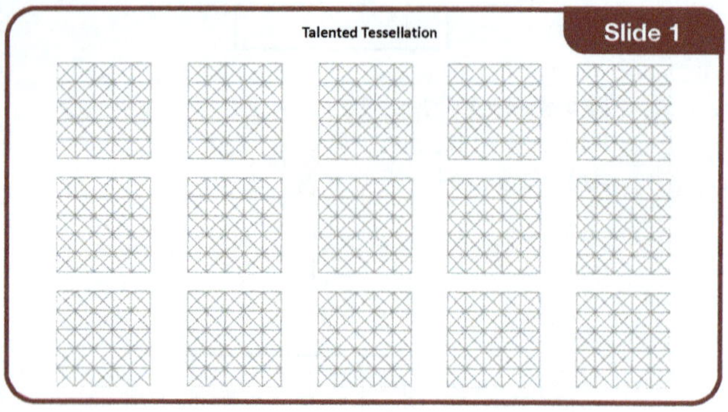

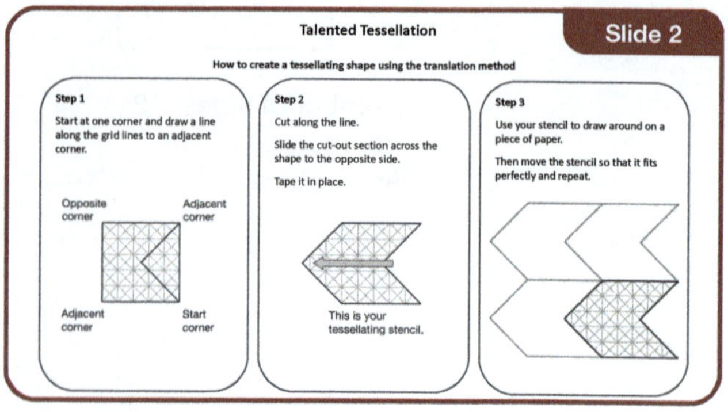

Print slides 1 and 2 and distribute to children. You will need one copy of slide 1 per child, plus a few extras. You will need one copy of slide 2 per child (or pairs could share).

Slides 3, 4 and 5

Reveal slides 3–5 one at a time, allowing time for discussion and observations.

For each slide, allow children to create the tessellating shapes and repeating patterns for themselves, using the 5 cm by 5 cm grids from slide 1. Sticky tape is the best way to connect the shapes together; glue would affect the perimeter and area due to the overlaps required.

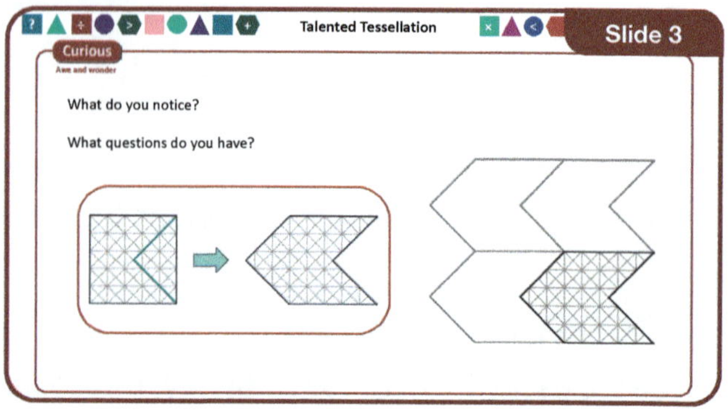

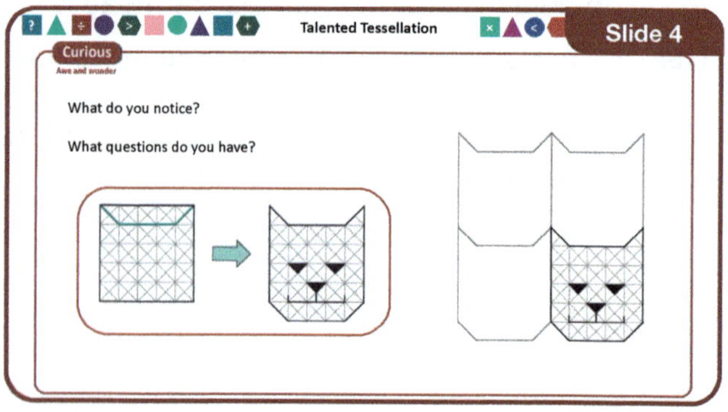

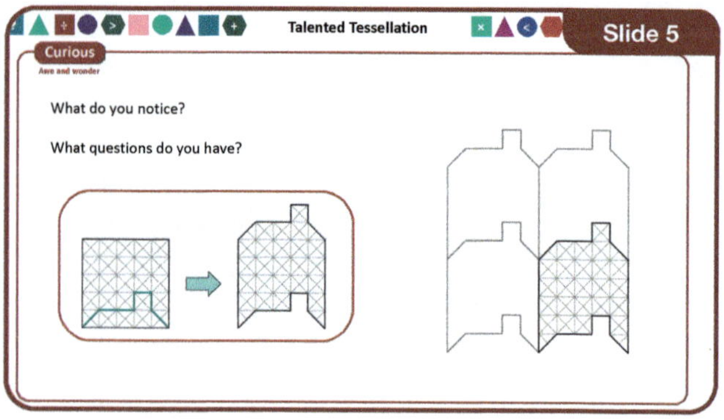

Slide 6

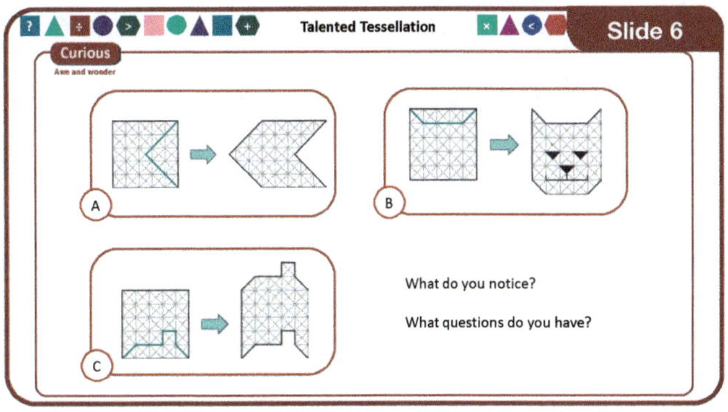

Display slide 6. Allow children to share what they are noticing or wondering, now that they have had a go at reproducing each of the stencils and patterns.

- I can see that the shapes all fit together.
- I can see that they all start with the same square grid.
- I have noticed that we move the cut-out shape to the opposite side every time.
- I can see that we always cut along the lines.
- This reminds me of translating shapes on a coordinate grid.

Now ask children to follow the instructions on their print-out of slide 2 to make a simple tessellating stencil of their own and check that it does tessellate without leaving any gaps. Tell the children that, at this stage, they should only cut along the straight lines provided, including the diagonals.

⚠ Children may not be familiar with the vocabulary of tessellating and translating. Emphasise this language when children are modelling and link it directly to their actions when creating the shapes and patterns.

Develop and deepen

Encourage children to use the words in the word bank to develop and deepen their thoughts and questions.

perimeter	**area**	**centimetre (cm)**
number of sides	**tessellate**	**translate**
centimetre squared (cm²)	**consistent**	**different**

⚠ Make sure children understand the difference between perimeter and area.

Possible lines of enquiry:

- I wonder whether I could make a more complicated shape that would still tessellate.

Possible lines of enquiry linked to the word bank:

- I wonder if the perimeter of each shape is the same or different.
- I wonder if the area of each shape is the same or different.
- I wonder if the perimeter / area changes as the number of sides increases.
- I wonder how I could make a shape with a longer perimeter than these, but so the shape still tessellates.
- I wonder if I could make a shape where the perimeter and the area match.
- I wonder whether there is a limit to how long the perimeter could be.

Possible lines of enquiry that children may raise now, but should be held back until the **Create** task:

- I wonder what would happen if the cut-out shapes were attached to the adjacent sides rather than the opposite sides. Would the shape still tessellate?
- I wonder if I could make a tessellating shape by translating cut-out shapes from more than one side.

Consider whether / how to collate and display all the suggested questions, and whether / how to steer children towards selecting an appropriately challenging question.

Pupil Resources for Full and Express Pathways

Distribute the **Pupil Resources** for **Talented Tessellation** to the children to help support them in their exploration of the different tasks. Use either of the **Curious Pupil Resources** depending on whether you are following the **Full Pathway** or the **Express Pathway**. The **Pupil Resources** for the remaining 5Cs are common to both pathways.

To provide additional flexibility in the classroom, each of the **Pupil Resources** is also provided in the **Teacher's Presentation** PowerPoint file (see slides 7–10), so you can display these at any time if you wish.

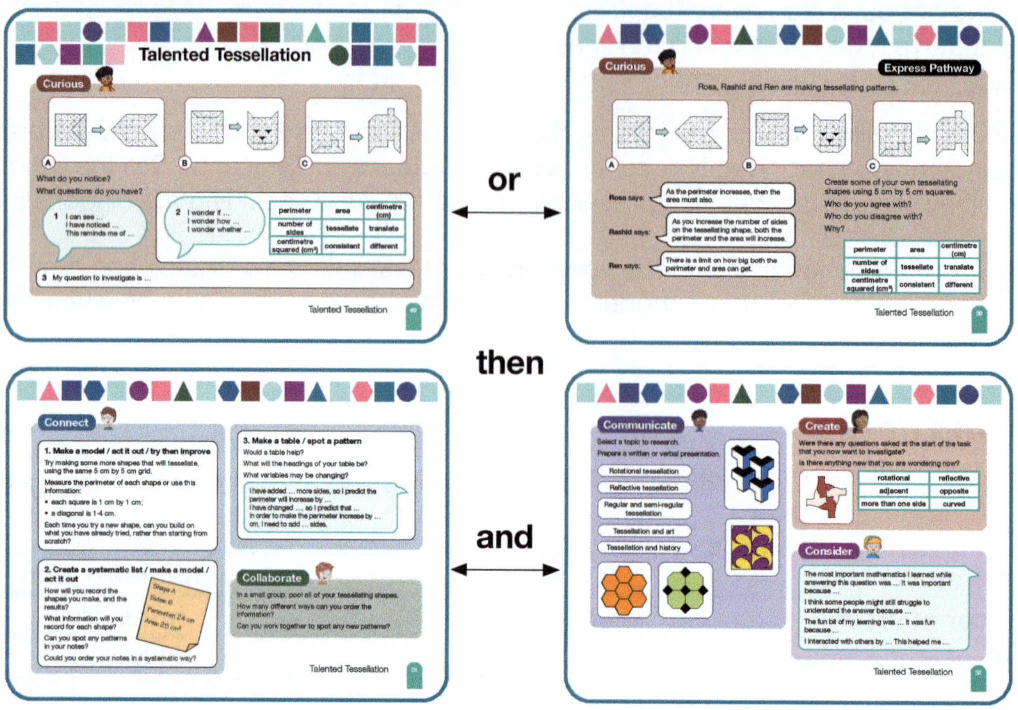

Express Pathway

If you are following the **Express Pathway**, spend a few minutes discussing the **Curious Pupil Resource** with the children. (This is slide 8, so you could display it if you wish.) Use some of the prompts on the previous page to help.

Three tasks involving different strategies are suggested for children to make connections.

1. Make a model / act it out / try then improve

Additional resources required: scissors, sticky tape, rulers

It would be helpful for children to start by creating a range of different tessellating shapes and checking their perimeters. Provide more print-outs of the grids from slide 1, if needed.

- *What shapes will you try out to help you investigate your chosen question?*

The perimeter of each shape can be measured using a ruler. Alternatively, share with children that each diagonal of a square is approximately 1·4 cm in length.

- *Each time you make a new shape, are you starting from scratch? Or are you looking at ways to improve on what you have already tried?*

2. Create a systematic list / make a model / act it out

Additional resources required: scissors, sticky tape, rulers, sticky notes

This task could be used as a follow-up to the previous **Connect** task, with children recording details of the shapes they have already made and then creating more to give them more data.

- *How will you record your attempts and the results?*
- *What are the important features that may or may not be changing?* (Number of sides, perimeter and area.)
- *How could you label your shapes, to help you keep track of them?* (A, B, C, or a name that describes the shape, like 'arrow'.)

Children could record the key information about each shape on sticky notes.

- *Can you see any patterns in the notes you have made?*
- *What about if you reorder them?*
- *How could you order them in a systematic way?* (By number of sides or by length of perimeter.)
- *Can you see any patterns now?*

3. Make a table / spot a pattern

- *How could a table help you in your explorations?* (It would keep the information organised and make it easier to spot patterns.)
- *What do you think the headings of the table should be?*

Slide 11

Print slide 11 and distribute to any children who are less confident about drawing their own table.

Ask children to record the key details about the shapes they have made so far. If they have already done some thinking about a systematic order for the shapes, they should follow this when recording in the table.

- *Can you see any patterns in your table?*

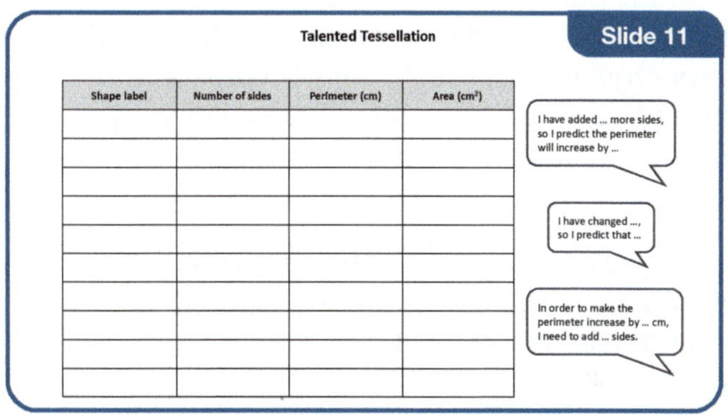

If needed, encourage children to redraw their tables and reorder the information to make the patterns clearer.

Shape label	Number of sides	Perimeter (cm)	Area (cm^2)
A (arrow)	6	24 cm	25 cm^2
B (cat)	8	21·6 cm	25 cm^2
C (house)	14	25·6 cm	25 cm^2

- *What if you reorder the information? Placing them in order of number of sides? Increasing perimeter?*
- *Why can't you reorder using area?*

Once the shapes have been recorded in their existing forms, children could adapt these, increasing the number of sides in an attempt to extend the perimeter.

- *Can you predict what will happen?*
- *How will the perimeter change? By how much?*
- *How will the area change? Why?*
- *How can you make sure that the new shape still tessellates?*
- *What is the maximum perimeter you can make?*
- *Is there a limit to the length of the perimeter for a tessellating shape?*

Collaborate

Refer to the **Collaborate** task of the **Pupil Resource**.

This task would work well with children working as a group of three or four.

Each child could create and explore their own tessellating shapes, recording the shape name / label, number of sides, perimeter and area onto a group table. The groups can then reorder their information in different ways to see if they can spot any patterns and make predictions.

They could be challenged to find tessellating shapes with perimeters that fit between their existing shapes.

- *How are you going to agree on the best way to order the information in the table?*
- *Do you need to agree a common way to name / label your shapes?*
- *Does any of the information need to be double-checked by someone else in the group?*

Communicate

Refer to the **Communicate** task of the **Pupil Resource**.

Slide 12

Display slide 12.

This task provides an opportunity for children to explore the beauty of mathematics. Tessellating patterns have a long historical background and have been used widely in different cultures. Children could choose an aspect of tessellation to research and create a written or verbal presentation, for example: rotational tessellation; reflective tessellation; regular and semi-regular tessellation; different artists who have used tessellation; use of tessellation by different cultures throughout history.

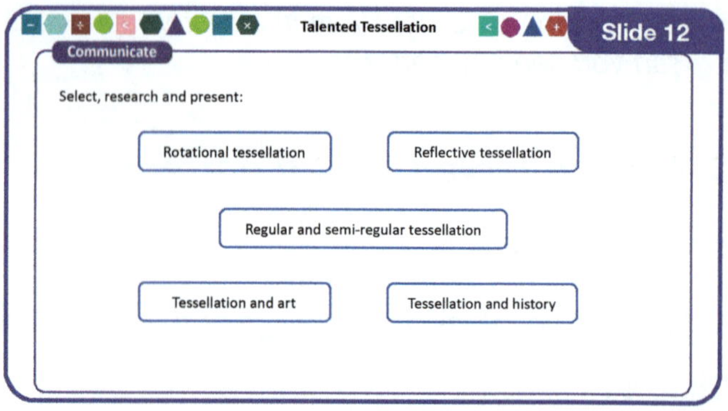

Slide 13

Display slide 13, to show children some simple examples of rotational, reflective, regular and semi-regular tessellation.

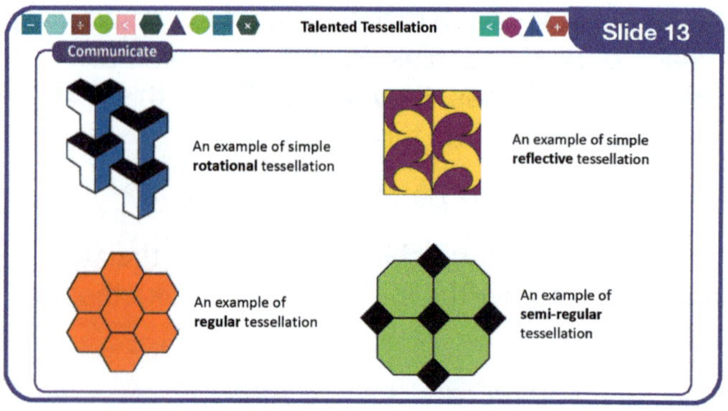

Refer to the **Create** task of the **Pupil Resource**.

Slide 14

Display slide 14.

- *Were there any questions asked at the start of the task that you now want to investigate?*
- *Is there anything new that you are wondering now?*

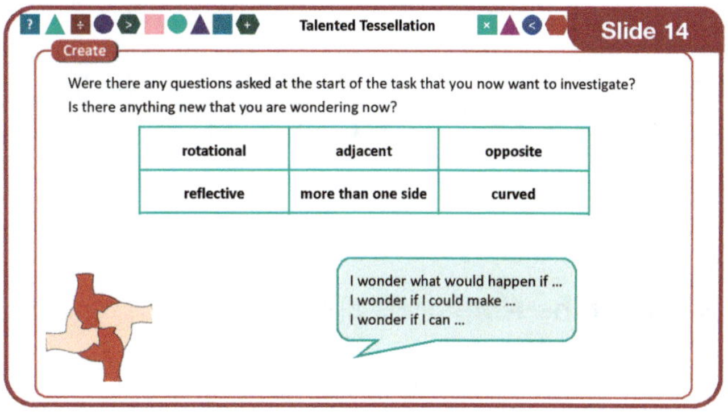

Children can use the word bank to prompt further questions of their own.

rotational	**adjacent**	**opposite**
reflective	**more than one side**	**curved**

- I wonder what would happen if the cut-out shapes were attached to the adjacent sides rather than the opposite sides. Would the shape still tessellate?
- I wonder if I could make a tessellating shape by translating cut-out shapes from more than one side.
- I wonder if I could make rotational tessellations.
- I wonder if I can cut curved lines to make a tessellating shape.

Refer to the **Consider** task of the **Pupil Resource**.

Slide 15

Display slide 15.

Encourage children to think about the strategies that they have used. The prompts below may help get children started.

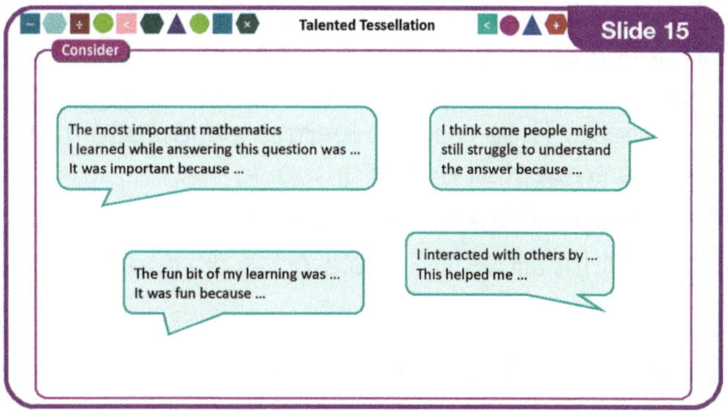

Cognitive learning domain

- The most important mathematics I learned while answering this question was … It was important because …
- I think some people might still struggle to understand the answer because …
- The trickiest part of solving my question was …

Affective domain

- The fun bit of my learning was … It was fun because …
- I said 'aha' when … I said it because …
- I was surprised about … because …

Metacognitive domain

- I interacted with others by … This helped me …
- Here are some tips I would give others for working well in a group: …
- I learned … about working in a group.

Battle of the Nets

Year 6

Introduction

Battle of the Nets gives children an opportunity to visualise and create nets of cubes and square cuboids. During the **Curious** stage, they are encouraged to think about the similarities and differences between these shapes and use this knowledge to predict whether they will have the same number of possible nets.

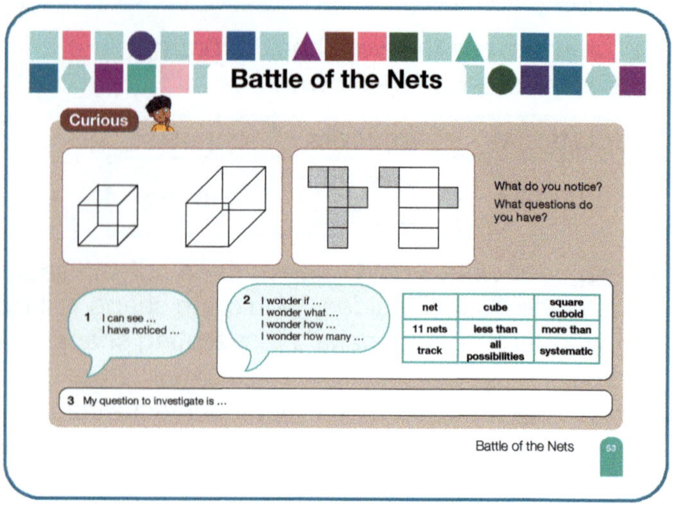

When making models or drawing diagrams, children should strive for accuracy, using appropriate measuring equipment. They will use precise mathematical language to describe their work. To confirm that they have found all possible nets for a square cuboid, children should work in a systematic way and consider how this means they can be efficient in the choices they make.

This task promotes greater-depth learning in a range of ways, supporting children to identify and articulate underlying mathematical structures. For example, in the **Collaborate** stage, greater depth is promoted by asking children to offer suggestions, listen carefully to others and build upon their own ideas. They will also learn to encourage and support others, and be able to accept it if their ideas are rejected by the group.

This **Teacher Guide** is supported by photocopiable **Pupil Resources** and **Teacher's Presentation** PowerPoint slides for whole-class or group display.

National Curriculum

Battle of the Nets covers these Year 6 National Curriculum objectives.

Geometry: properties of shapes

- Draw 2D shapes using given dimensions and angles.
- Recognise, describe and build simple 3D shapes, including making nets.
- *Pupils draw shapes and nets accurately, using measuring tools and conventional markings and labels for lines and angles.* [non-statutory guidance]

The 6C cycle

The grid on the right shows the relative weighting given to each of the 6 Cs in **Battle of the Nets**.

Curious	★★
Connect	★★
Collaborate	★★★
Communicate	★★
Create	★★
Consider	★★

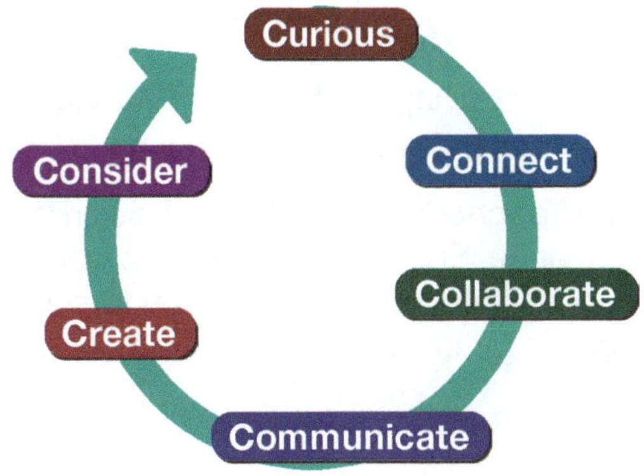

Problem-solving heuristics

The grid shows which problem-solving strategies children are likely to use when investigating the **Connect** tasks in **Battle of the Nets**. These strategies are further exemplified in the **Connect** stage below (pages 148–149).

Draw a diagram	✓
Make a table / spot a pattern	
Make a model / act it out	✓
Try then improve	✓
Create a systematic list	✓
Solve a simpler but related problem	
Work backwards	

Key mathematical ideas

In order to get the most out of **Battle of the Nets**, children need to understand what nets are, and it would be useful if they have already explored building 3D shapes and nets for familiar 3D shapes.

The task assumes prior experience of exploring and discovering that there are 11 different possible nets for a cube. (See notes in **Awe and wonder** below for what to do if this is not the case.)

Children may assume that because square cuboids have the same number of faces as a cube, there must be the same number of possible nets. During the task, they will discover that there are actually a lot more (29 in total).

Discussions will be needed about whether all the nets children find are truly unique, or if any are just a reflection or rotation of one they have already found, and how they can be sure.

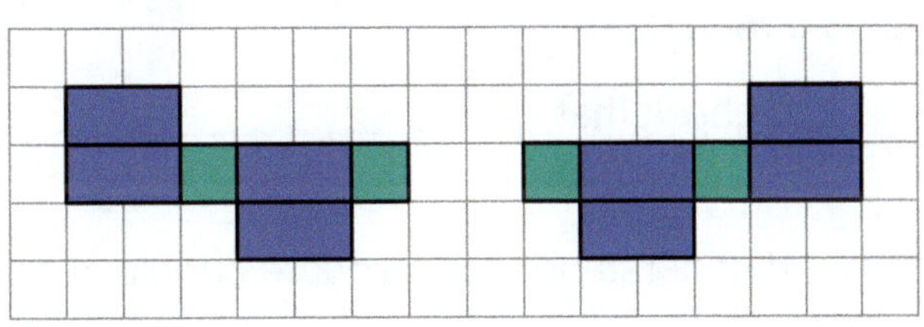

Square cuboids (cuboids with at least two square faces) have been chosen for exploration, although children may choose to extend their investigations to looking at other types of cuboid (that is ones with no square faces).

When representing nets, children should be encouraged to use measuring equipment to help them to accurately draw the 2D faces.

Answers to the tasks are given on page 152.

Teaching guidance

If you are using the **Express Pathway**, please turn to pages 147–151.

Teaching guidance for the **Full 6C Pathway** is given below.

The **Teacher's Presentation** PowerPoint file has slides specifically mentioned in this teaching guidance. Some of the slides include notes (within the PowerPoint file) providing 'stumble support' suggestions, that may help you to support children in their problem-solving thinking and enable access for children who are less confident.

Awe and wonder

Slide 1

Display slide 1.

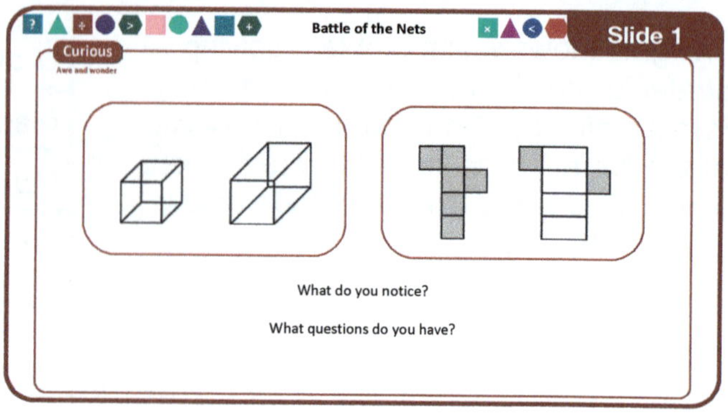

Ask children to share what they are noticing or wondering.

- I can see two 3D shapes.
- I have noticed that the nets in the second image represent the shapes in the first image.
- I can see that the cuboid is larger than the cube.
- I have noticed that the cuboid has two square faces.
- I have noticed that the shapes have some things in common. They both have some square faces. Each face has 4 right angles. They both have 8 vertices and 12 edges.
- I have noticed that the nets are similar. They both have at least two square faces. The nets are arranged in the same way. The difference is that in the cuboid, four of the faces are rectangles.

Slide 2

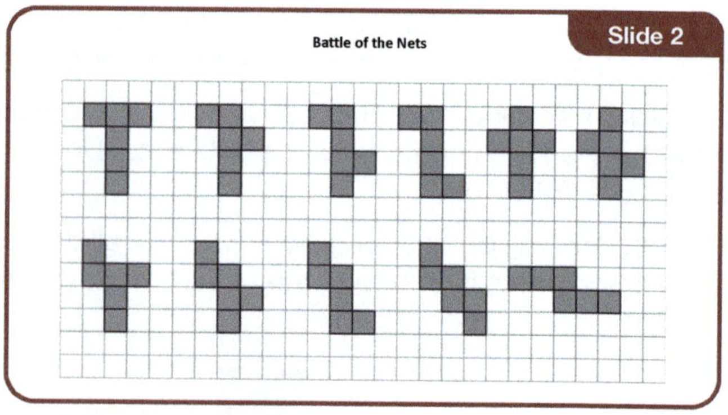

If children do not have prior experience of exploring and discovering that there are 11 different nets for a cube, print slide 2 and distribute to children.

- *Which nets will definitely make a cube?*
- *Are there any nets that you are unsure of?*
- *What could you do to test your thoughts?*

Children could cut out the nets and check that they each make a cube.

⚠ Ensure that children know and understand that a square cuboid is a special type of cuboid that has at least two square faces.

Develop and deepen

Encourage children to use the words in the word bank to develop and deepen their thoughts and questions.

net	**cube**	**square cuboid**
11 nets	**less than**	**more than**
track	**all possibilities**	**systematic**

Possible lines of enquiry:

- I wonder if the nets given will match the two 3D shapes.
- I wonder what other shapes I can draw a net for.

Possible lines of enquiry linked to the word bank:

- I wonder if both shapes have the same number of possible nets.
- I wonder if there are more or less than 11 nets for the square cuboid.
- I wonder how we can keep track of all of the nets to make sure that we have found all possibilities.

Possible lines of enquiry that children may raise now, but should be held back until the **Create** task:

- I wonder how many nets a different cuboid would have where the faces do not include squares.
- I wonder how many possible nets there are for other 3D shapes (for example, square-based pyramids, triangular prisms or octahedrons).

Consider whether / how to collate and display all the suggested questions, and whether / how to steer children towards selecting an appropriately challenging question.

Pupil Resources for Full and Express Pathways

Distribute the **Pupil Resources** for **Battle of the Nets** to the children to help support them in their exploration of the different tasks. Use either of the **Curious Pupil Resources** depending on whether you are following the **Full Pathway** or the **Express Pathway**. The **Pupil Resources** for the remaining 5Cs are common to both pathways.

To provide additional flexibility in the classroom, each of the **Pupil Resources** is also provided in the **Teacher's Presentation** PowerPoint file (see slides 3–6), so you can display these at any time if you wish.

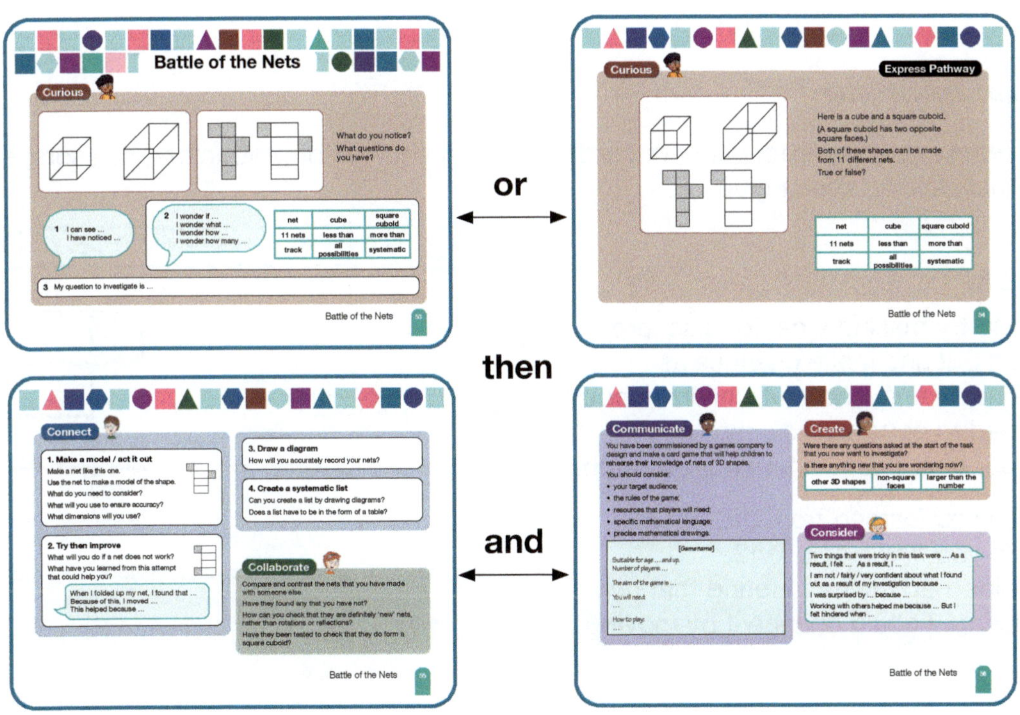

Express Pathway

If you are following the **Express Pathway**, spend a few minutes discussing the **Curious Pupil Resource** with the children. (This is slide 4, so you could display it if you wish.) Use some of the prompts on the previous page to help.

Four tasks involving different strategies are suggested for children to make connections.

1. Make a model / act it out

Additional resources required: paper or card, rulers, set squares or protractors, scissors, squared paper (optional), sticky tape or glue

Slide 7

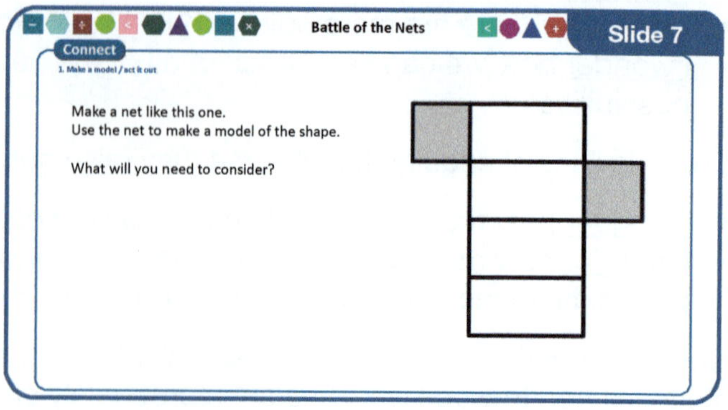

Display slide 7.

Ask children to start by making a copy of the net on the slide and making a model of it to test that it works.

- *What do you need to consider?*
- *What will you use to ensure accuracy?*
- *What dimensions will you use for your net?*
- *Why would dimensions such as 2 cm × 3 cm × 4 cm not be suitable?*

Children should draw their net accurately using measuring equipment such as rulers and protractors or set squares to ensure right angles (or they could use squared paper).

- *How will you ensure your cuboid stays together?*
- *Have you built shapes from a net before?*

If children are building nets that they want to glue together, they will need to include tabs. Decide whether to leave children to find this out for themselves or whether to model doing this.

When children have successfully made their first net and used it to construct a square cuboid, ask them to create different nets that will make the same shape.

2. Try then improve

Additional resources required: paper or card, rulers, set squares or protractors, scissors, squared paper (optional), sticky tape or glue

Slide 8

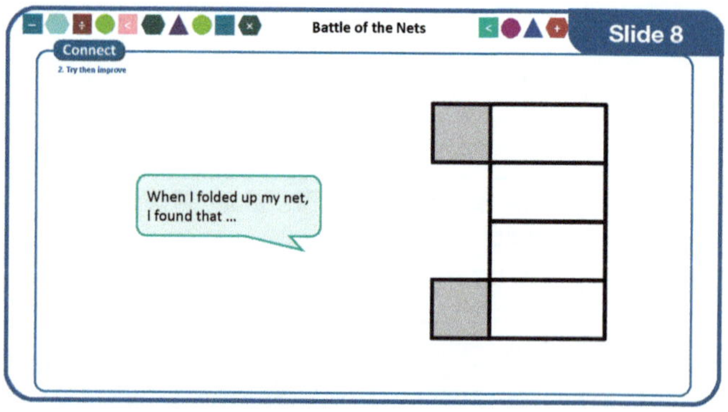

Display slide 8.

Children could try making a net for a square cuboid, test it out and make changes as needed. If a net is unsuccessful, children should be encouraged to consider how they might adapt the net rather than starting from scratch.

- *What have you learned from this net that could help you find a solution?*
- When I folded up my net, I found that two faces overlap. I could change it by moving …

3. Draw a diagram

Additional resources required: squared paper, coloured pencils, rulers (optional), set squares or protractors (optional)

Once children have created their nets, these should be recorded on paper. They could do this by shading squares on squared paper, or they could draw their nets on plain paper, using measuring equipment such as rulers and set squares to ensure accuracy.

4. Create a systematic list

Additional resources required: squared paper, coloured pencils

Slide 9

Display slide 9.

Dependent on their chosen line of enquiry, children may be exploring if they can find all of the different nets for a square cuboid. They should be encouraged to work systematically to do this.

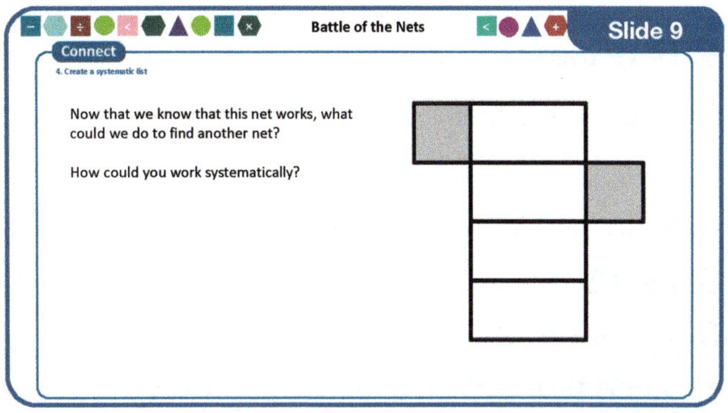

Slide 10

Display slide 10.

It may be useful to start by asking children to look at the 'list' of cube nets and discuss what they notice about the way that the solutions have been presented.

- I have noticed that in the top row, the 'base' and the 'lid' have been put into different positions, with the middle section remaining the same.

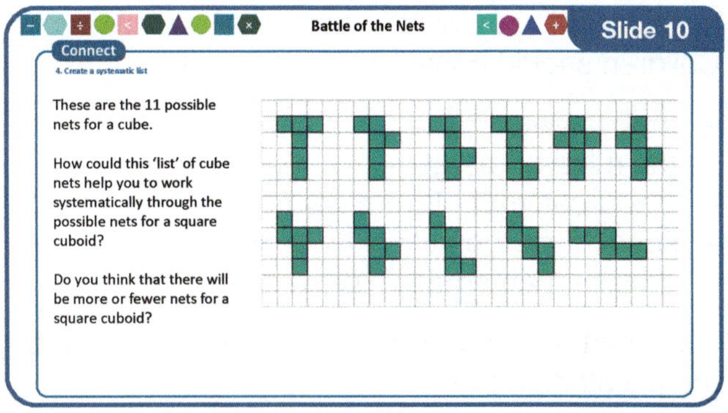

Children could then be encouraged to imitate this to help them begin to find as many nets as possible for a square cuboid. It may be helpful for them to shade the square faces in a different colour to the rectangular faces, to help spot rotations / reflections.

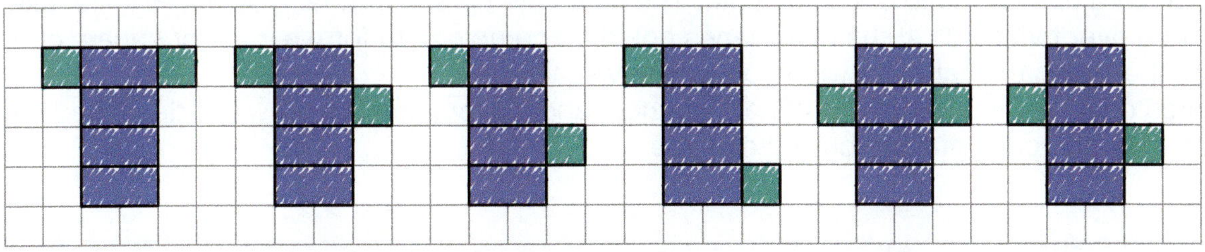

- I have found all of the different nets if the rectangular faces are arranged one above the other. I found some reflections but I discounted these.

Collaborate

Refer to the **Collaborate** task of the **Pupil Resource**.

Slide 11

Display slide 11.

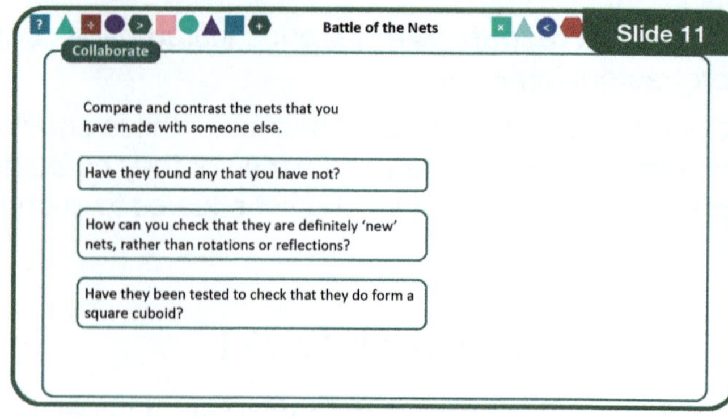

This task would provide a good opportunity for children to work in pairs or small groups. They could compare and contrast the nets that they have made to check that they have not made any replications.

- *Has your partner / anyone in your group found any nets that you have not?*
- *How can you check that they are definitely 'new' nets, rather than rotations or reflections?*
- *Have these additional nets been tested to check that they do form a square cuboid?*

Children could collaborate further when creating their game as detailed in the **Communicate** task.

Communicate

Refer to the **Communicate** task of the **Pupil Resource**.

Slide 12

Display slide 12.

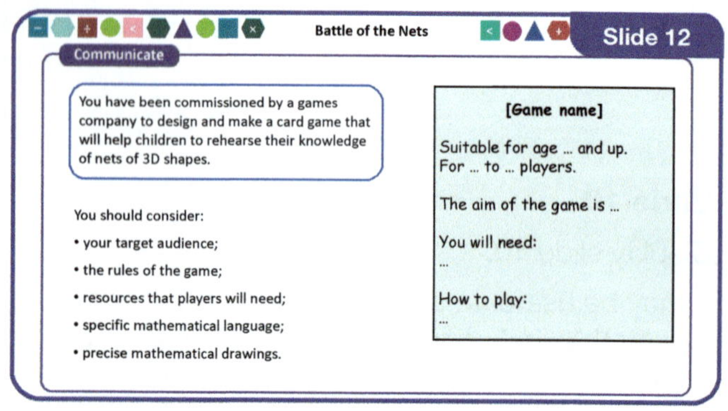

This task provides an opportunity for the children to design a card game with the purpose being to identify nets of 3D shapes. Children should consider:

- their target audience;
- the rules of the game;
- resources that players will need;
- specific mathematical language;
- precise mathematical drawings.

If children are struggling to come up with an idea, you could share the following examples to give them a starting point.

- Examples and non-examples of nets of cubes or square cuboids, or a mixture of both, are drawn on cards. Cards are placed face down. One player turns over a card. If this card has a correct example of a net on it, they collect the card. If it has a non-example, then they turn the card back over and play moves on. The winner is the player who collects the most cards.
- Each card has an image of half of a net. The cards are spread out, face down. Players take turns to turn over two cards. If the two images could be combined to form a cube or square cuboid net, they keep the pair. This will involve clear visualisation.
- Cards could have descriptions of nets on them and players need to draw the described net, or draw the net, cut it out and build the shape.

Refer to the Create task of the **Pupil Resource**.

Slide 13

Display slide 13.

- *Were there any questions asked at the start of the task that you now want to investigate?*
- *Is there anything new that you are wondering now?*

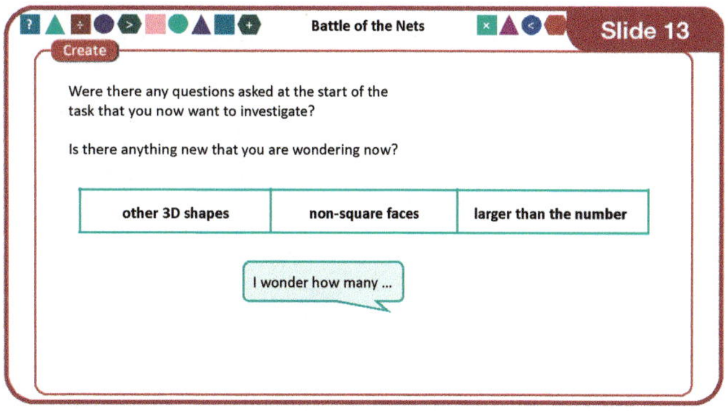

Children can use the word bank to prompt further questions of their own.

other 3D shapes	**non-square faces**	**larger than the number**

- I wonder how many nets a different cuboid would have where the faces do not include squares.
- I wonder how many possible nets there are for other 3D shapes (for example, square-based pyramids, triangular prisms or octahedrons).
- I wonder if the number of nets is larger than the number of sides of one of the 2D faces. For example, will a triangular-based pyramid have more than three nets?

Refer to the Consider task of the **Pupil Resource**.

Slide 14

Display slide 14.

Encourage children to think about the strategies that they have used. The prompts below may help get children started.

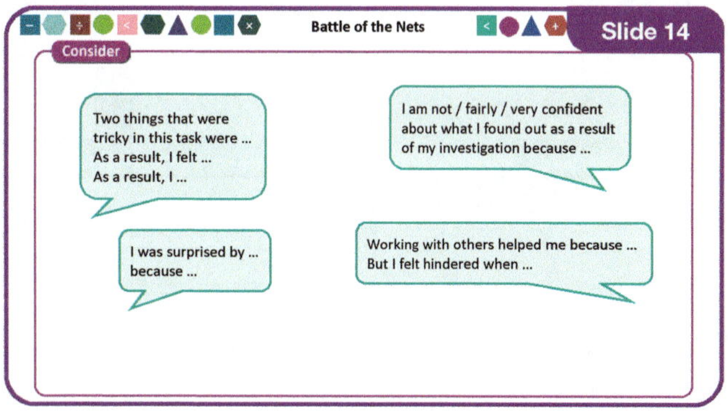

Cognitive learning domain

- Two things that were tricky in this task were … As a result, I felt … As a result, I …
- I think that … might make this task difficult for someone because …
- I am not / fairly / very confident about what I found out as a result of my investigation because …

Affective domain

- I found … frustrating. I think that this was due to …
- I was surprised by … because …
- I said 'aha' when … because … As a result, I …

Metacognitive domain

- Working with others helped me because … But I felt hindered when …
- The cooperation strategies that helped me to learn were …
- Next time, I will …

Answers

There are 29 possible nets for a square cuboid.

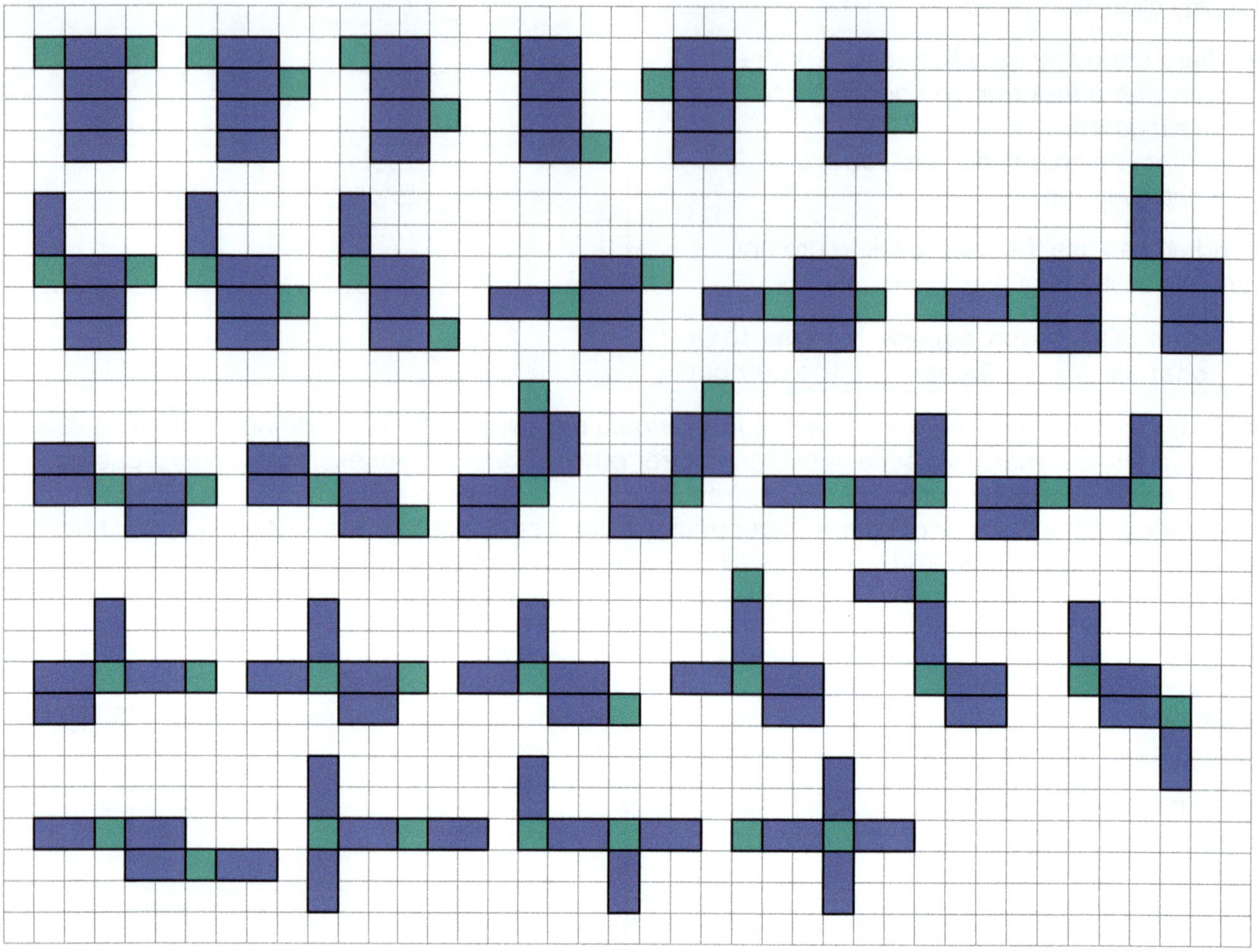

For a non-square cuboid (for example, one with dimensions of 4 cm × 5 cm × 6 cm), there are 54 possible nets.

Numbers of possible nets for other 3D shapes:

- octahedron: 11 nets
- triangular prism: 9 nets
- square-based pyramid: 6 nets
- triangular-based pyramid: 2 nets

Distorted Data

Year 5

Year 6

Introduction

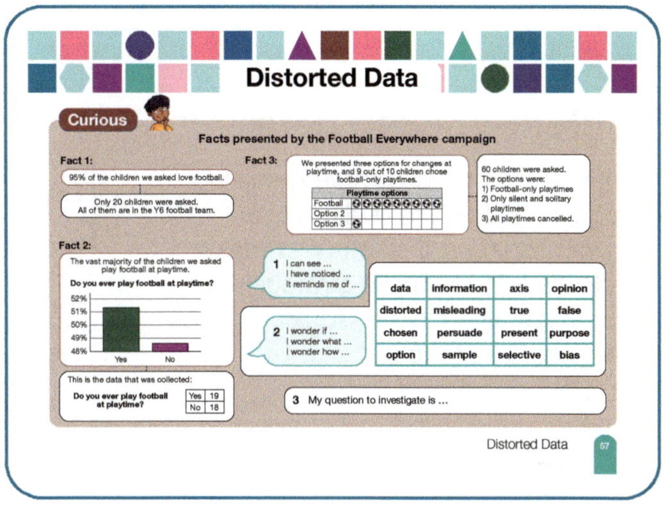

Distorted Data provides an opportunity for children to deepen their experiences of reading and interpreting data by being faced with a collection of purposefully misleading data. Throughout this task, children will need to interrogate the given data asking a variety of appropriate, mathematical questions and assessing the quality of how the information has been presented.

Children will then have the chance to demonstrate what they have learned, by collecting their own accurate data sets but then presenting them purposefully to favour a particular outcome. This will allow children to also include fractions and percentages, and Year 6 children may also want to consider presenting their data in pie charts or including the mean average.

This task promotes greater-depth learning in a range of ways, supporting children to identify and articulate underlying mathematical structures. For example, in the **Communicate** stage, greater depth is promoted by asking children to develop clear, accurate and well-organised communication, with ideas connected and flowing from one to the next, and illustrated using diagrams or models for clarification as needed.

This **Teacher Guide** is supported by photocopiable **Pupil Resources** and **Teacher's Presentation** PowerPoint slides for whole-class or group display.

National Curriculum

Distorted Data covers these Year 5 and Year 6 National Curriculum objectives.

Year 5 Statistics

- Complete, read and interpret information in tables, including timetables.
- *Pupils begin to decide which representations of data are most appropriate and why.* [non-statutory guidance]

Year 6 Statistics

- Interpret and construct pie charts and line graphs and use these to solve problems.
- Calculate and interpret the mean as an average.

The 6C cycle

The grid on the right shows the relative weighting given to each of the 6 Cs in **Distorted Data**.

Curious	★★★
Connect	★★
Collaborate	★★★
Communicate	★★★
Create	★
Consider	★★

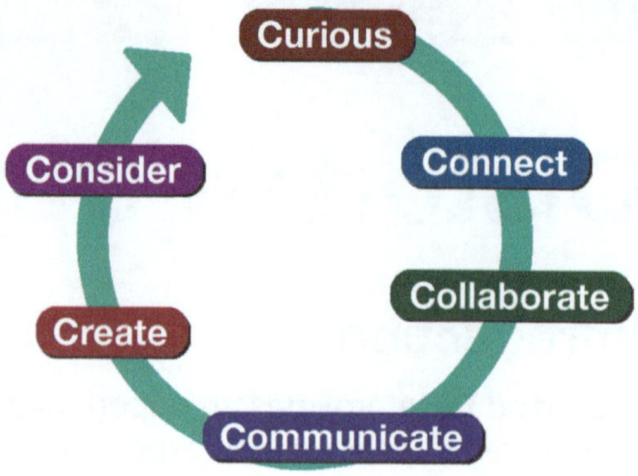

Problem-solving heuristics

The grid shows which problem-solving strategies children are likely to use when investigating the Connect tasks in **Distorted Data**. These strategies are further exemplified in the Connect stage below (pages 157–158).

Draw a diagram	✓
Make a table / spot a pattern	
Make a model / act it out	
Try then improve	✓
Create a systematic list	
Solve a simpler but related problem	
Work backwards	✓

Key mathematical ideas

Data is used in many different ways in everyday life such as persuading people what to buy, what to watch, who to vote for and lifestyle choices. This task provides children with an opportunity to explore 'fake news' which is a term that children in Upper Key Stage 2 are likely to have come across. Because of this, it is important for children to become statistically literate. They need to understand how to read, interpret and question data and headline messages. Does the data tell the whole story? How accurate is it? Who has collected and produced the data? What was their purpose for collecting the data?

There are a number of methods that can be employed when creating misleading / distorted statistics. In this task, three key ones have been exemplified, through the presentation of misleading headline data, followed by the revealing of additional information.

- Purposeful and selective bias (asking only a biased population who have a vested interest in the outcome): Fact 1 claims that '95% of the children we asked love football'; but only 15 children in the school were surveyed, and those children are all on the football team.
- Misleading graphics (manipulating the axis scales to emphasise or hide a gap): Fact 2 claims that 'The vast majority of the children we asked play football at playtimes', and the accompanying bar chart appears to back this up; but when the scale on the *y*-axis is changed it becomes clear that there is actually only a very small difference between the percentage of people who answered 'Yes' and 'No'.
- Faulty polling (phrasing questions in such a manner that people answer in a predictable way): Fact 3 claims that 'We presented three options for changes at playtime, and 9 out of 10 children chose football-only playtimes'; but the other two options were worded in such a way that they would appeal to very few children.

Due to the nature of this task, no answers are provided.

Teaching guidance

If you are using the **Express Pathway**, please turn to pages 157–160.

Teaching guidance for the **Full 6C Pathway** is given below.

The **Teacher's Presentation** PowerPoint file has slides specifically mentioned in this teaching guidance. Some of the slides include notes (within the PowerPoint file) providing 'stumble support' suggestions, that may help you to support children in their problem-solving thinking and enable access for children who are less confident.

⚠ Some children may need to gain more experience of the use of data. If needed, share some real-life examples and discuss how the facts presented may not tell the whole story.

Awe and wonder

Reveal slides 1–2 one at a time, allowing time for discussion and observations as prompted below.

Slide 1

Display slide 1.

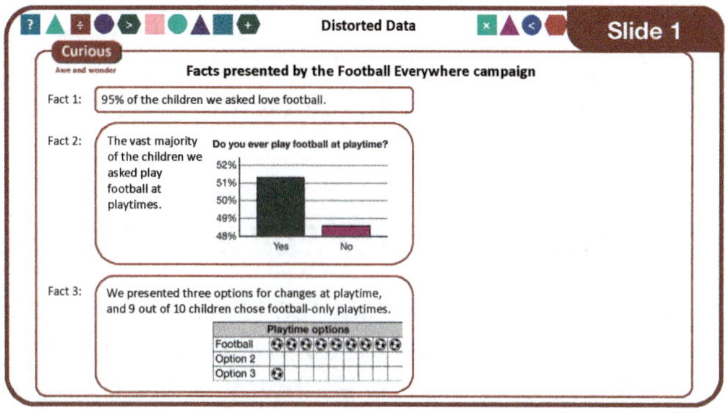

- I have noticed that most children they asked love football, because 95% agreed.
- I can see from the graph that most children said that they have played football at playtime.
- It reminds me of an advert that is trying to get you to buy something.
- I can see that football was the most popular option for playtimes, because 90% chose it.

Slide 2

Display slide 2.

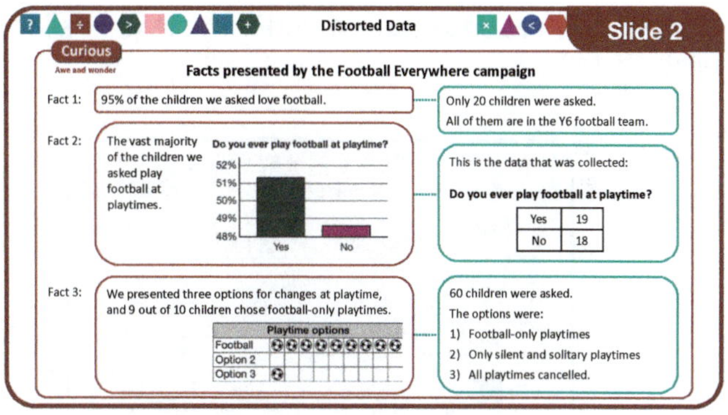

- I can see that, for Fact 1, they only asked the Year 6 football team. That doesn't seem fair; of course they love football!
- I have noticed that, for Fact 2, the data table shows that the numbers were actually really close. The scale chosen for the *y*-axis on the graph makes it looks as though the gap is much bigger.
- I can see that, for Fact 3, the other two options for playtimes sounded really bad, so it's not surprising that hardly anyone chose them!

Develop and deepen

Encourage children to use the words in the word bank to develop and deepen their thoughts and questions.

data	**information**	**axis**	**opinion**
distorted	**misleading**	**true**	**false**
chosen	**persuade**	**present**	**purpose**
option	**sample**	**selective**	**bias**

Possible lines of enquiry linked to the word bank:

- I wonder how the axis on a graph can be used to present misleading data.
- I wonder how information can be used to change people's opinions.
- I wonder how the people who are chosen for a survey can affect the data collected and presented.
- I wonder what the data for Fact 1 would be if they had asked more children and not just those on the football team.
- I wonder if children would have agreed to the football-only option in Fact 3 if there was an option to make no changes to playtimes at all.

Possible lines of enquiry that children may raise now, but should be held back until the **Create** task:

- I wonder what other strategies can be used to change how data looks to the reader.

Consider whether / how to collate and display all the suggested questions, and whether / how to steer children towards selecting an appropriately challenging question.

⚠ Some children may need help to distinguish between the concept of untrue (falsified) data and distorted data (carefully presented to favour a cause). They may need time to unpick that none of the three facts are untrue, but have been presented in a way that promotes the campaign.

Pupil Resources for Full and Express Pathways

Distribute the **Pupil Resources** for **Distorted Data** to the children to help support them in their exploration of the different tasks. Use either of the **Curious Pupil Resources** depending on whether you are following the **Full Pathway** or the **Express Pathway**. The **Pupil Resources** for the remaining 5Cs are common to both pathways.

To provide additional flexibility in the classroom, each of the **Pupil Resources** is also provided in the **Teacher's Presentation** PowerPoint file (see slides 3–6), so you can display these at any time if you wish.

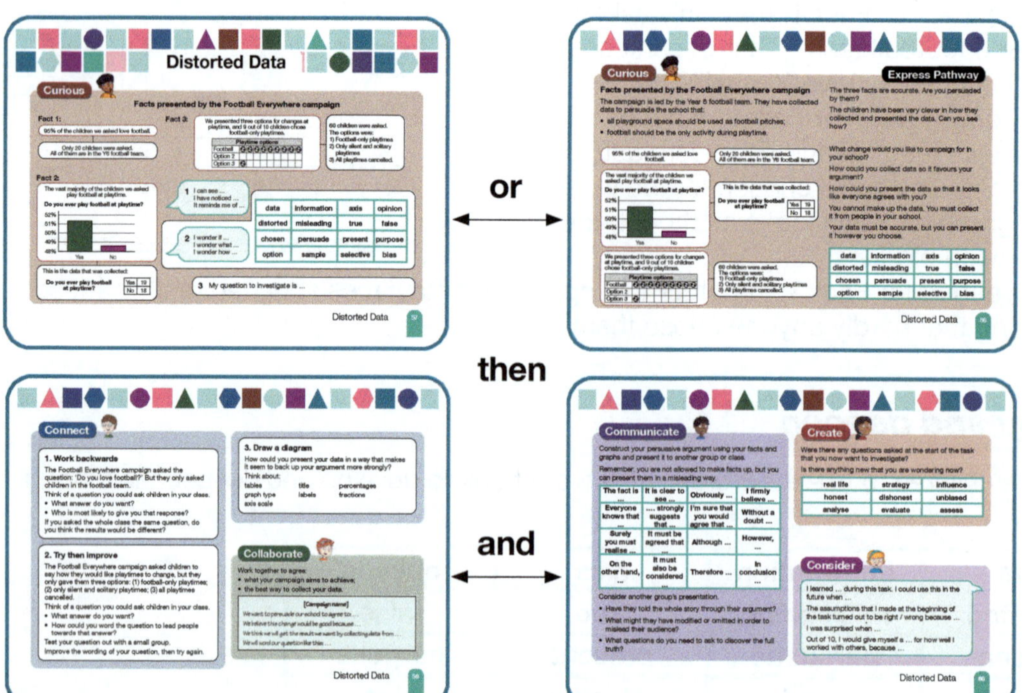

Express Pathway

If you are following the **Express Pathway**, spend a few minutes discussing the **Curious Pupil Resource** with the children. (This is slide 4, so you could display it if you wish.) Use some of the prompts on the previous page to help.

Three tasks involving different strategies are suggested for children to make connections.

Before beginning the **Connect** stage, ask children to form pairs or small groups with whom they will work for the rest of the task. Allow some time for them to discuss and agree something they would like to campaign for. They should then have a brief initial conversation about how they might collect data to support their campaign.

1. Work backwards

Slide 7

Display slide 7.

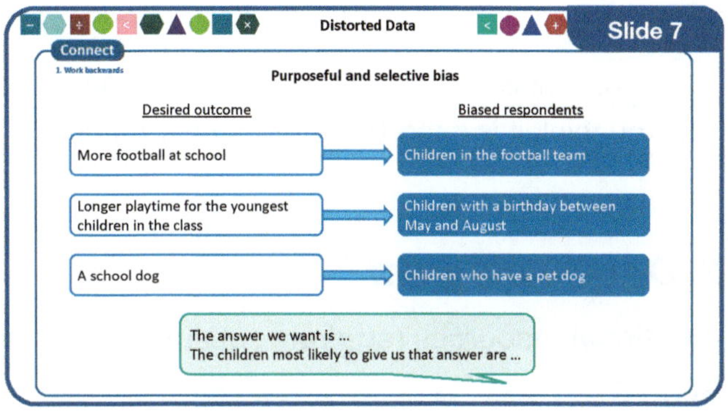

This task is designed to get children thinking about purposeful and selective bias, and the possibility of ensuring favourable data by targeting key groups when conducting their survey.

- *The Football Everywhere campaign targeted the question 'Do you love football?' to the football team only. Why do you think they did this?*

Ask children to think of a desired outcome, then work backwards to think about how it could be obtained by selecting biased respondents.

Examples could include:

- longer playtimes for the younger children in the class (target the question at children whose birthday is between May and August);
- getting a school dog (target the question at children who have pet dogs).

Give children the opportunity to carry out their one-question survey with their targeted group in the class. If possible, allow them to repeat the survey, but this time asking the whole class. This will allow them to compare their results from targeted and non-targeted groups.

2. Try then improve

Slide 8

Display slide 8.

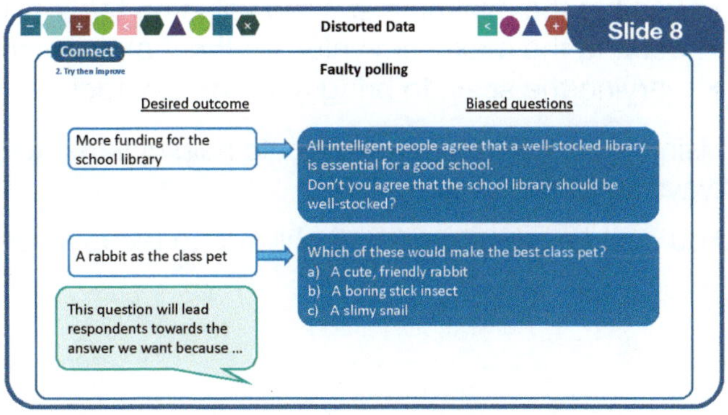

In this task, children should consider how they could manipulate the responses through the wording of the question or the response options.

- *The Football Everywhere campaign asked children how they would like playtimes to be changed, but the only options they gave were: (1) football-only playtimes; (2) only silent and solitary playtimes; and (3) all playtimes cancelled. Why do you think they did this?*

Ask children to think of a question and a desired outcome.

- *How could you phrase the question so that the majority of people will agree with you?*
- *Think about which option you want them to select. What alternative options could you present?*

Give children time to survey a random selection of children in the class, using the question they have written, and record the data in a table.

- *Did you get the result you wanted?*
- *How could you refine the wording of your question or the response options to get more people to agree with you?*

If possible, allow children to test out their improved question on another group of children to compare the two sets of data.

For example, '37·5% of children want a nature club' is less persuasive than 'A huge 87·5% of children agree that looking after nature is important.'

	Yes	No
Don't you think a nature club would be really good?	37·5%	62·5%
Do you think it is important to look after nature?	87·5%	12·5%

3. Draw a diagram

Additional resources required: squared paper, rulers, spreadsheet software (optional)

Slide 9

Display slide 9.

In this task, children will think about how the way data is presented visually can distort people's perception of it.

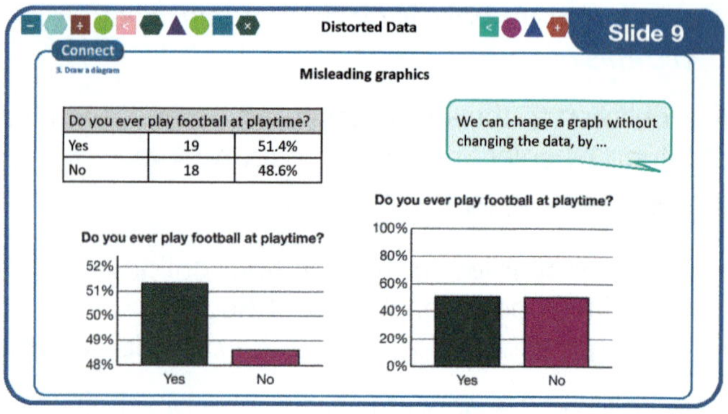

- *The Football Everywhere campaign chose a bar chart to represent the responses to the question: 'Do you ever play football at playtime?' But they chose a scale for the y-axis that makes it look as though lots more people answered 'Yes'. This is what the bar chart looks like when the scale goes from 0% to 100%.*

Ask children to either gather or make up some data, and then try out different ways to present it using graphs or charts. For example, they could adapt the axis on a bar chart to either emphasise the difference or similarities within data. This could include:

- varying the minimum starting value on the *y*-axis (for example, starting at zero or not);
- varying the maximum value on the *y*-axis which can impact the height of the bars;
- varying the scale to bring more or less focus to certain results within the data.

Using spreadsheet software could help children to 'see' how the data could be presented in different ways rapidly.

You could encourage Year 6 children to focus on constructing and adapting a pie chart.

Refer to the **Collaborate** task of the **Pupil Resource**.

Slide 10

Display slide 10.

Give children time to discuss and decide on their chosen campaign within their pairs or small groups, and plan out how they will collect their data.

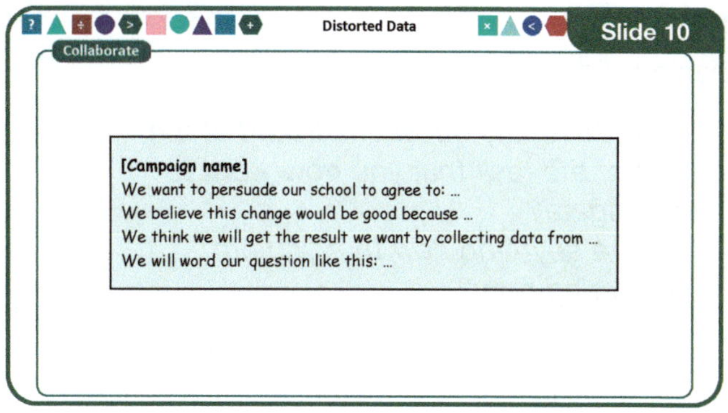

- *What change do you want people to agree to?*
- *What question will you ask?*
- *What response options will you give?*
- *Who will you collect data from?*

Then ask children to conduct their surveys and collect their data.

- *How will you present your data visually?*
- *What sort of tables, charts or graphs will you use?*
- *How will you manipulate the graphs to show what you want them to show, without changing the actual data?*

Refer to the **Communicate** task of the **Pupil Resource**.

Slide 11

Display slide 11.

This task provides an opportunity for the children to present a persuasive argument either in written or verbal form from two contrasting viewpoints.

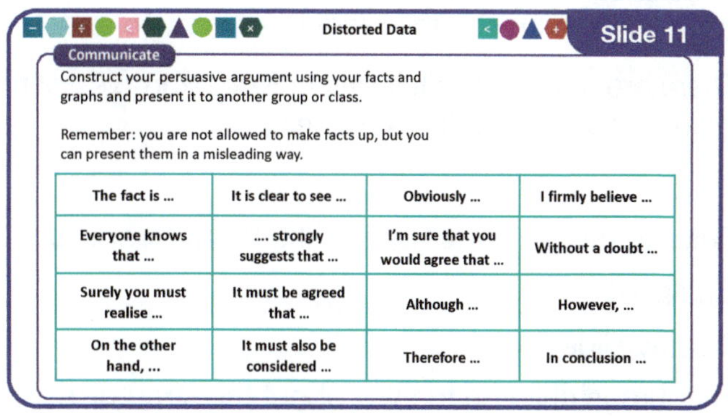

Firstly, children will need to use their distorted data to their advantage and prepare a presentation for another group or class.

The word bank could be useful in helping children with their persuasive writing.

Slide 12

Display slide 12.

Ask children to consider the argument presented by another group and attempt to expose the distorted data.

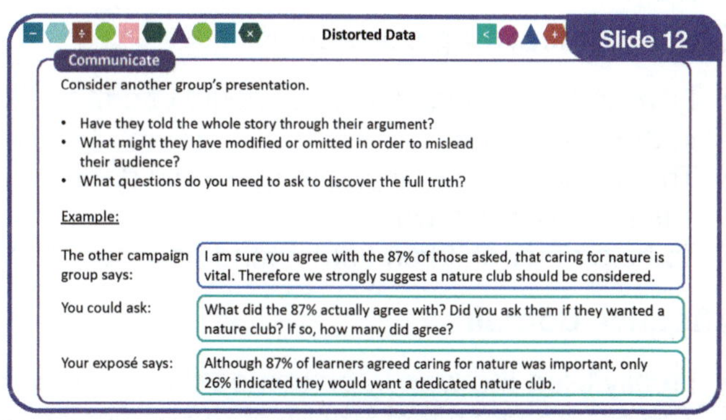

- *Have they told the whole story through their argument?*
- *What might they have modified or omitted in order to mislead their audience?*
- *What questions do you need to ask discover the full truth?*
- Some questions I could ask are ...

You may wish to give groups time to interview each other through a 'hot seat' and then write an exposé.

Create

Refer to the **Create** task of the **Pupil Resource**.

Slide 13

Display slide 13.

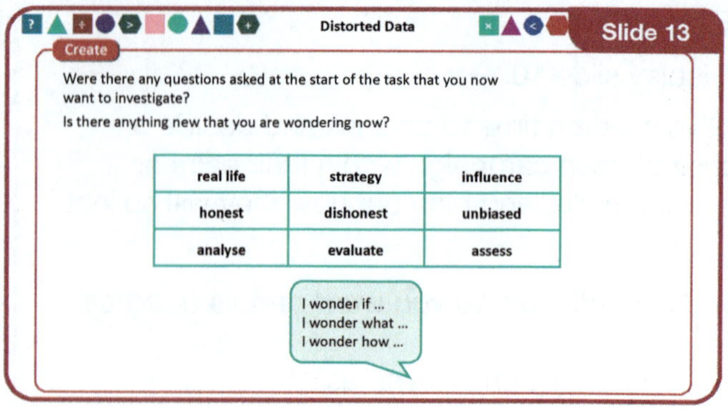

- *Were there any questions asked at the start of the task that you now want to investigate?*
- *Is there anything new that you are wondering now?*

Children can use the word bank to prompt further questions of their own.

real life	**strategy**	**influence**
honest	**dishonest**	**unbiased**
analyse	**evaluate**	**assess**

- I wonder if I could find real-life examples of where data has been presented in a dishonest way to support persuasive articles or campaigns.
- I wonder what other strategies can be used to change how data looks to the reader.
- I wonder how I can analyse the data I am looking at to find out if it is honest.
- I wonder how I could share what I've learned to ensure others know about how data can be used to influence the reader and how to assess whether it is unbiased.

There are many websites that collate examples of misleading data from advertising campaigns, including social media examples.

Consider

Refer to the **Consider** task of the **Pupil Resource**.

Slide 14

Display slide 14.

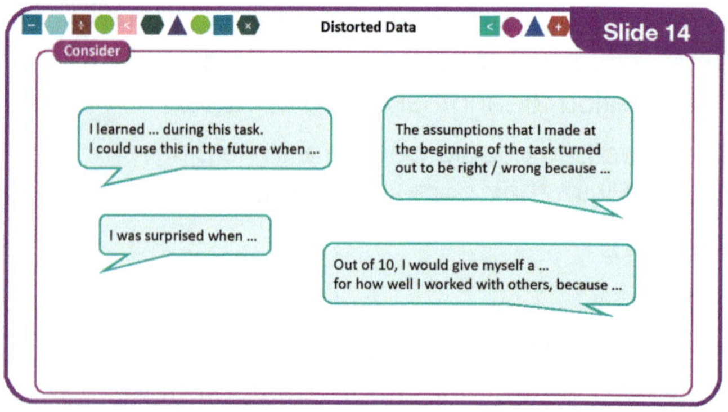

Encourage children to think about the strategies that they have used. The prompts below may help get children started.

Cognitive learning domain

- I learned … during this task. I could use this in the future when …
- The trickiest part of trying to make my data suit my argument was …
- To help other children understand how data can be misleading, I could …

Affective domain

- I really liked the way … presented their argument because …
- The assumptions that I made at the beginning of the task turned out to be right / wrong because …
- I was surprised when …

Metacognitive domain

- Out of 10, I would give myself a … for how well I worked with others, because …
- I think I could be a better group member next time by …
- If I had to do this task again, I would make sure that I …